Engineering Chemistry

Dr.T. Indumathi

Published by

Engineering Chemistry

ISBN 978-93-86638-57-1

Author

Dr.T. Indumathi

Bonfring

309, 2nd Floor, 5th Street Extension, Gandhipuram,
Coimbatore-641 012.
Tamilnadu, India.
E-mail: info@bonfring.org
Website: www.bonfring.org
Phone: 0422 4213231

Acknowledgement

I would like to express my gratitude to the many people who saw me through this book; to all those who provided support, talked things over, read, wrote, offered comments, allowed me to quote their remarks and assisted in the editing, proofreading and design.

I also wish to thank the management of Sri Krishna College of Engineering and Technology and PSG College of Technology, Coimbatore for the academic support and the facilities provided to carry out my text book work. I also extend my sincere thanks to the Principal, Head of the Department and all other teaching and non-teaching staff members for their valuable support throughout my text book writing.

I am grateful to my mother, my husband, my son and other family members for their continuous support and cooperation. My deemed acknowledgement to all friends who have helped in all phases of my text book for its timely completion. Last and not least: I beg forgiveness of all those who have been with me over the course of the years and whose names I have failed to mention."

Unit	Contents	Page No

UNIT I

CHEMISTRY IN EVERYDAY LIFE

1.1. Analgesic

An **analgesic** (also known as a **painkiller**) is any member of the group of drugs used to relieve pain (achieve analgesia). The word *analgesic* derives from Greek *an-* ("without") and *algos* ("pain").

Analgesic drugs act in various ways on the peripheral and central nervous systems; they include paracetamol (para-acetylaminophenol, also known in the US as acetaminophen), the non-steroidal anti-inflammatory drugs (NSAIDs) such as the salicylates, and opioid drugs such as morphine and opium. They are distinct from anaesthetics, which reversibly eliminate sensation.

Analgesic Can Be Divided Into two Main Groups Namely

1. Narcotic analgesics These drugs produce depression of the central nervous system and are two types.
   - Natural analgesics (eg. Morphine, codeine etc.).
   - Synthetic analgesics (eg. pethidine, methadone etc.).
2. Non-narcotic analgesics.

The drugs do not produce significant depression of the central nervous system, but it possesses antipyretic and anti-inflammatory property. Antipyretic agents reduce the elevated body temperature. It regulates the body temperature by cutaneous vasodilation and by excessive sweating.

Non-narcotic analgesics are classified as:

1. The Salicylic acid and its derivatives.

   eg.Asprin [non-steroidal anti-inflammatory drugs (NSAIDs)]
   Methyl salicylate, sodium salicylate, salicin etc.
2. Para-aminophenolderivatives.

   eg. Para paracetamol, phenacetin.
3. The Pyrazole derivatives.EgAnalgin.
4. Indolyl and Aryl acetic acid derivatives.

The exact mechanism of action of paracetamol/acetaminophen is uncertain, but it appears to be acting centrally rather than peripherally (in the brain rather than in nerve endings).

Aspirin and the other non-steroidal anti-inflammatory drugs (NSAIDs)inhibit cyclooxygenases, leading to a decrease in prostaglandin production. This reduces pain and also inflammation (in contrast to paracetamol and the opioids.

Paracetamol has few side effects and is regarded as safe, although intake above the recommended dose can lead to liver damage, which can be severe and life-threatening, and occasionally kidney damage.

NSAIDs predispose to peptic ulcers, renal failure, allergic reactions, and occasionally hearing loss, and they can increase the risk of haemorrhage by affecting platelet function.

The use of aspirin in children under 16 suffering from viral illness has been linked to **Reye's syndrome**, a rare but severe liver disorder.

1.2.   Antiseptics

Antiseptics are antimicrobial substances that are applied to living tissue or skin to reduce the possibility of infection, sepsis, or putrefaction.

Antiseptics are generally distinguished from *antibiotics* by the latter's ability to be transported through the lymphatic system to destroy bacteria within the body, and from *disinfectants*, which destroy microorganisms found on non-living objects.

Some antiseptics are true *germicides*, capable of destroying microbes (bacteriocidal), whilst others are bacteriostatic and only prevent or inhibit their growth.

Functionality

For the growth of bacteria, there must be a food supply, moisture, in most cases oxygen, and a certain minimum temperature.

Some Common Antiseptics

- **Alcohols**: Most commonly used is ethanol (60–90%), 1-propanol (60–70%) and 2-propanol/isopropanol (70–80%) or mixtures of these alcohols. They are commonly referred to as "surgical alcohol" Used to disinfect the skin before injections are given, often along with iodine (tincture of iodine).

- **Quaternary ammonium compounds**: It is also known as *Quats* or *QAC's* include the chemicals benzalkonium chloride (BAC), cetyltrimethylammonium bromide (CTMB), cetylpyridinium chloride (Cetrim, CPC) and benzethonium chloride (BZT). Benzalkonium chloride is used in some pre-operative skin disinfectants (conc. 0.05–0.5%) and antiseptic towels.

* **Boric acid:** It is used in suppositories to treat yeast infections of the vagina, in eyewashes, and as an antiviral to shorten the duration of cold sore attacks.
* **Iodine:** It is usually used in an alcoholic solution (called tincture of iodine) or as Lugol's iodine solution as a pre- and post-operative antiseptic.
* **Manuka Honey:** It is used as a medical device for use in wounds and burns.

1.3. Antacid

An **antacid** is a substance, generally a base or basic salt, which neutralises stomach acidity.

Mechanism of Action

Antacids perform a neutralisation reaction, increasing the pH to reduce acidity in the stomach. When gastric hydrochloric acid reaches the nerves in the gastrointestinal mucosa, they signal pain to the central nervous system. This happens when these nerves are exposed, as in peptic ulcers.

Some Well-Known Chemicals used as Antacid is,

* $NaHCO_3$ and/or $KHCO_3$
* $Al(OH)_3$ and $Mg(OH)_2$
* $CaCO_3$ $MgCO_3$
* Milk of Magnesia – $Mg(OH)_2$
* $Al(OH)_3$

1.4. Disinfectant

Disinfectants are substances that are applied to non-living objects to destroy microorganisms that are living on the objects. Disinfection does not necessarily kill all microorganisms, especially nonresistant bacterial spores. It is less effective than *sterilisation*. Disinfectants are different from other antimicrobial agents such as *antibiotics*, which destroy microorganisms within the body, and *antiseptics*, which destroy microorganisms on living tissue. Disinfectants are frequently used in hospitals, dental surgeries, kitchens, and bathrooms to kill infectious organisms.

Types of Disinfectants

1. Air Disinfectants

Air disinfectants are typically chemical substances capable of disinfecting microorganisms suspended in the air. An air disinfectant must be dispersed either as an aerosol or vapour at a sufficient concentration in the air to cause the number of viable infectious microorganisms to be significantly reduced.

2. Alcohols

Alcohols, usually ethanol or isopropanol, are sometimes used as a disinfectant. They are non-corrosive, but can be a fire hazard. They also have limited residual activity due to evaporation

3. Aldehydes

Aldehydes, such as formaldehyde and glutaraldehyde, have a wide microbiocidal activity and are sporicidal and fungicidal. They are partly inactivated by organic matter and have slight residual activity.

4. Oxidizing Agents

Oxidizing agents act by oxidizing the cell membrane of microorganisms, which results in a loss of structure and leads to cell lysis and death. Chlorine and oxygen are strong oxidizers

a) Sodium hypochlorite is very commonly used. Common household bleach is a sodium hypochlorite solution and is used in the home to disinfect drains, toilets, and other surfaces. In more dilute form, it is used in swimming pools, and in still more dilute form, it is used in drinking water.

b) Chloramine is often used in drinking water treatment.

c) Chlorine dioxide is used as an advanced disinfectant for drinking water to reduce waterborne diseases.

d) Hydrogen peroxide is used in hospitals to disinfect surfaces and it is used in solution alone or in combination with other chemicals as a high-level disinfectant. Hydrogen peroxide vapour is used as a medical sterilant and as room disinfectant.

Peracetic acid, Lactic acid,Performic acid and Potassium permanganate are also used as **disinfectants.**

5. Phenolics

Phenolics are active ingredients in some household disinfectants. They are also found in some mouthwashes and in disinfectant soap and hand washes. **Phenol is probably the oldest known disinfectant.** It is rather corrosive to the skin and sometimes toxic to sensitive people. Impure preparations of phenol were originally made from coal tar, and these contained low concentrations of other aromatic hydrocarbons including benzene, which is an IARC Group 1 carcinogen.

- Chloroxylenol is the principal ingredient in Dettol, a household disinfectant and antiseptic.

- Hexachlorophene is a phenolic that was once used as a germicidal additive to some household products but was banned due to suspected harmful effects.

- Thymol, derived from the herb thyme, is the active ingredient in some "broad spectrum" disinfectants that bears ecological claims.

1.5. Artificial Sweetening Agents

Artificial Sweetener, any synthetically produced, intensely sweet substance developed for use in reduced-calorie or diet foods and beverages. Unlike natural sugars, artificial sweeteners do not promote decay, and they contribute few if any calories in the foods they sweeten. Major sweeteners include aspartame, saccharin, the cyclamates, and sucralose. Sweeteners are usually made from the fruit or sap of plants, but can also be made from any other part of the plant, or all of it. Some sweeteners are made from starch, with the use of enzymes. Sweeteners made by animals, especially insects, are put in their own section as they can come from more than one part of plants. The sap of some species is concentrated to make sweeteners, usually through drying or boiling. Monosaccharide and disaccharide are sweet in taste. However, there are some compounds which are sweeter than the sugar. These synthetically developed compounds are called as artificial sweeteners.

Examples of artificial sweeteners are:

- Saccharin.
- Aspartame.
- Sucralose.
- Cyclamates.

1.6. Sugar Substitutes

A sugar substitute is a food additive that duplicates the effect of sugar in taste, usually with less food energy. Some sugar substitutes are natural and some are synthetic. Those that are not natural are, in general, called artificial sweeteners.

An important class of sugar substitutes is known as high-intensity sweeteners. These are compounds with a sweetness that is many times that of sucrose, common table sugar.

Six intensely-sweet sugar substitutes have been approved for use. They are aspartame, sucralose, saccharin and cyclamate. The majority of sugar substitutes approved for food use is artificially-synthesized compounds. Some bulk natural sugar substitutes are known, including sorbitol and Xylitol.

For example, xylose is converted to Xylitol, lactose to lactitol, and glucose to sorbitol. Still other natural substitutes are known but are yet to gain official approval for food use. Artificial sweeteners are among the most controversial food additives due to allegations of adverse health effects. These allegations include dermatological problems, headaches, mood variations, behaviour changes, respiratory difficulties, seizures, and cancer.

1.7. Properties

1. It is an odourless, white crystalline powder that is derived from the two amino acids aspartic acid and phenylalanine.
2. It is about 200 times as sweet as sugar and can be used as a tabletop sweetener or in frozen desserts, gelatins, beverages, and chewing gum.

3. When cooked or stored at high temperatures, aspartame breaks down into its constituent amino acids. This makes aspartame undesirable as a baking sweetener. It is more stable in somewhat acidic conditions, such as in soft drinks. Though it does not have a bitter aftertaste like saccharin, it may not taste exactly like sugar. When eaten, aspartame is metabolised into its original amino acids.

4. Aspartame cannot be considered a "non-caloric" sweetener since it is broken down in the digestive tract into its components which are absorbed and metabolised. These components, aspartic acid, phenylalanine, and methanol, account for the 4 Calories per gram energy rating of aspartame.

5. However, since the substance is about 180 times sweeter than sugar, very little needs to be used in foods and beverages to achieve a satisfactory degree of sweetness.

6. Aspartame cannot be used in cooked or baked foods since it breaks down into its components and loses its sweeteningpower.

7. Aspartame is not stable in acidic conditions or when heated.

1.8. Saccharin

Saccharin is a synthetic organic compound that is 200-100 times sweeter as cane sugar. The sodium or calcium salt of saccharin is widely used as a diet sweetener.

It is 300 to 500 times as sweet as sugar (sucrose) and is often used to improve the taste of toothpaste, dietary foods, and dietary beverages.

Saccharin, synthetic, white, crystalline powder, of formula $C_6H_4CONHSO_2$, which in its pure state is 550 times as sweeter as sugarcane. In its commercial form, saccharin is estimated to have a sweetening power 375 times that of sugar. It is prepared from toluene.

When moderately large amounts of saccharin dissolved in water, the solution has a bitter taste, the sweetness being evident only in dilute solutions.

Saccharin is not digested by the body and has no food value. It is used in place of sugar by persons suffering from diabetes (Diabetes Mellitus) and persons with weight-reducing diets, for the psychological purpose of satisfying a taste for sweetness.

1.9.   Sucralose

- Sucralose is a chlorinated sugar that is about 600 times as sweet as sugar. It is produced from sucrose when three chlorine atoms replace three hydroxyl groups.

- It is used in beverages, frozen desserts, chewing gum, baked goods, and other foods.
- Unlike other artificial sweeteners, it is stable when heated and can, therefore, be used in baked and fried foods.
- About 15% of sucralose is absorbed by the body and most of it passes out of the body unchanged
- The only artificial sweetener made from sugar is sucrose.

1.10. Cyclamates

- Three similar compounds, namely sodium cyclamate, calcium cyclamate, and cyclamic acid, are collectively referred to as the "cyclamates".
- They are about thirty times sweeter than sugar and are chemically more stable than saccharin or aspartame. Some of a cyclamate dose is excreted by the body unchanged, but some are converted to a cyclohexylamine.

1.11. Non-Sugar Sweeteners

Some non-sugar sweeteners are polyols, also known as "sugar alcohols." These are, in general, less sweet than sucrose, but have similar bulk properties and can be used in a wide range of food products.

1.12. Reasons for Use

- **To assist in weight loss.**
- **Dental care:** Sugar substitutes are tooth-friendly, as they are not fermented by the microflora of the dental plaque.
- **Diabetes mellitus :** People with diabetes have difficulty regulating their blood sugar levels.
- **To reduce cost** : Many sugar substitutes are cheaper than sugar.
- **Avoiding processed foods:** Individuals may opt to substitute refined white sugar with less-processed sugars, such as fruit juice or maple syrup.

1.13. Food Preservatives

Artificial preservatives are a group of chemical substances added to food, sprayed on the outside of food, or added to certain medications to retard spoilage, discoloration, or contamination by bacteria and other disease organisms.

Preservatives Should Have the Following Properties

1. It should not react with the ingredient or with thecontainer.
2. Must be non-toxic.
3. It should be soluble in any medium.

Artificial Food Preservatives

Chemical food preservatives are synthesised artificially that are

- Benzoates(sodium benzoate, benzoic acid).
- Nitrites (sodium nitrites), Sulphites.
- Sorbets (sodium sorbets, potassium sorbets).

Benzoate & Benzoic Acid

These two compounds are related because sodium benzoate produces benzoic acid when it is dissolved in water, It has antimicrobial properties and naturally found in plums, ripe cloves, apples etc.

Sodium benzoate, salt of benzoic acid is preferred over benzoic acid in many food applications because it is 180 times more soluble in water. It has a marked pH effect (ie) lowers the pH, more effective it is. So it will work if the food product has a pH below 4.5.It is used in jams, beverages, dressing salads, icings.

Sorbate (Sodium Sorbets, Potassium Sorbate)

Sorbic acid is the compound with antimicrobial properties, due to the difference insolubility. Potassium sorbate is effective upto pH 6.5. If pH decreases its effectiveness also decreases. In wine processing sorbate is used to prevent refermentation. In many food products, sorbate and benzoates are used together to provide greater protection against a wide variety of micro-organism.

Nitrites (Sodium Nitrites)

Prevents the growth of bacteria, particularly Clostridium botulinum (bacterium responsible for botulism), in meat or smoked fish.

Sulphites - Dried fruits, wines and fruit juices.

Antimicrobial agents	Function
Benzoates.	inhibits the growth of moulds, yeasts, and bacteria in acidic drinks and liquids, including fruit juice, vinegar, sparkling drinks and soft drinks
Sodium benzoate	with a pH below 3.6, including salad dressings, carbonated drinks, fruit juices, and Oriental food sauces such as soy sauce and duck sauce
sorbate	Prevents the growth of moulds, yeasts, and fungi in foods or beverages.
Nitrites	Prevents the growth of bacteria, particularly Clostridium botulinum (bacterium responsible for botulism), in meat or smoked fish.
Antioxidants	**Function**
Butylatedhydroxy toluene	inhibits the growth of yeasts and fungi in beer and wines, and preserves meats, dried potato products, and dried fruits
Sulfites.	Used in fats, oils, shortening, and similar products
Vitamin C	Slows oxidation of fresh-cut fruits and vegetables used to fortify breakfast cereals and pet foods.
VitaminE.	Prevents browning of freshapples, peaches, and other fruits.

Some preservatives aid in manufacturing. eg Anti-caking agents prevent liquids from boiling or from froth when they are bottled.

Thickening and stabilising agents which are added to give a certain consistency and texture to food such as pectin in ajam, locust bean gum in ice cream to prevent the formation of anice crystal.

Emulsifier: Food containing fats and water Lecithin in chocolates, mono and diglycerides in bakery products.

Colouring agent: Caramel in soft drinks.

Chelating agent: citric acid, EDTA.

1.14. Application of Chemical in Daily Life

Two significant contributions made by chemistry towards our lifestyle is the protection of our health and hygiene. The purification of our water by the process of chlorination, dental cleanliness from toothpaste, sterilisation, control and the cure of disease are all a part of daily living discovered and created by chemistry. These discoveries make up a major component for the protection of our health and hygiene.

Hard water is healthy as it consists of calcium, Ca^{2+}, and Mg^{2+} are in need in the formation of bones and are important in the clotting of the blood and regulating the heartbeat. Mg^{2+} are needed for making proteins and for passing impulses along the nerve cells. The Chlorination of water for drinking makes the water safe because it rids it from the diseases which can be transmitted through water.

$$Cl_2 + H_2O \longrightarrow HCl + HOCl \quad \text{(Hypochlorous acid)}$$

One aim of chemistry is to maintain dental cleanliness. Bacterial infection of tooth structure is called dental caries. The bacteria convert the sugar in our diet, particularly sucrose, to a glue to stick themselves to the tooth surface. Acids such as acetic, propanoic and lactic are also produced. These acids cause the calcium phosphate in the tooth enamel to dissolve.

$$Ca_3(PO_4)_2 + 4H^+ \rightarrow Ca(H_2PO_4)_2$$

The aim of toothpaste is to clean and polish the teeth. It basically consists of a mild abrasive, usually powdered calcium carbonate and a small amount of soap. The solids are suspended in a liquid, usually glycerol, and dyes and flavourings are added.

Sterilisation is the destruction of bacteria and is an essential part in the fight against thedisease. Milk is partially sterilised by heating it to about $80^\circ C$ and maintaining the temperature for about 16 seconds. This process is called pasteurisation and ensures that all potentially harmful bacteria destroyed within the milk. Chemists have produced many bacteria-killing chemicals and the use of these has saved many lives.

The first of these chemicals was a dilute solution of **carbolic acid, C_6H_5OH**. In1985, a surgeon named **Joseph Lister** began sterilising surgical wounds with this solution. Two main solutions used to kill bacteria in the home are antiseptics and disinfectants. Disinfectants are stronger than antiseptics and are too harsh, irritant or toxic to be used on the skin. It is usually used to clean or sterilise objects such as surgical instruments. Antiseptics are used on the skin, either before surgery or on cuts and bruises, and prevent sepsis, which is the destruction of the tissues by bacteria.

One aim of medicine is to control thedisease with the use of chemicals. Chemicals or drugs are made from natural causes such as a plant or animal extract, or made in the laboratory. The successful synthetic drug is acetylsalicylicacid or more known as aspirin. Aspirin is anti-inflammatory, meaning it reduces swelling, redness, etc. It is also known as analgesics as it is apain killer.

Another aim of medicine is to cure disease. The use of chemical agents to destroy infectious organisms, or diseases, without destroying the host is called chemotherapy. Very significant types of drugs related to chemotherapy are drugs containing sulphur drugs. Sulpha drugs are dyes that are lethal to harmful microorganisms. Protonsil was the first sulfa drug. It breaks down in the body to sulfanilamide, which is effective in destroying streptococci.

There are many contributions made by the chemistry that deal with ourlives. The simple health and hygiene of a human, in particular, consisted of many hours or hard and dedicated work and research by chemists.

The purification of our water, dental cleanliness by toothpaste, the sterilisation of a cut, the control of disease by aspirin, and the cure of disease by chemotherapy have all contributed to our livelihood and our health.

1.15. Water

Characteristic of Water

Water quality is determined by physical, chemical and microbiological properties of water. These water quality characteristics throughout the world are characterisedby wide variability. Therefore, the quality of natural water sources used for different purposes should be established in terms of the specific water-quality parameters that most affect the possible use of water.

Physical Characteristics of Water

Physical characteristics include colour, taste, odour etc., are determined by senses of touch, sight, smell and taste.

Colour

Colour in water is primarily a concern for water quality. Coloured water gives the appearance of being unfit to drink, even though the water may be perfectly safe for public use. On the other hand, colour can indicate the presence of organic substances, such as algae or humic compounds.

More recently, colour has been used as a quantitative assessment of the presence of potentially hazardous or toxic organic materials in water.

Colour is measured by 'tintometer' and expressed in mg/lit in platinum-cobalt scale or Hazen units.

Turbidity

Turbidity is a measure of the water clarity. It depends on upon the amount of material suspended in water.

Suspended materials can be the soil particles and other substances. Turbidity often increases sharply during a rainfall season. Turbidity also depends on upon the stream flow and velocity.

Turbidity is measured by using theturbid meter and Nephelometer. A light scattering property of water is used to know the turbidity. Turbidity is measured in nephelometric turbidity units, (NTU).

Chemical Characteristics of Water

1. Acidity

Acidity is the "measure of the ability of water to neutralise bases (or) it is the tendency to donate H^+ in order to neutralise the basic anions". Acidity is usually expressed in terms of pH.

The potential of the hydrogen ion solution, i.e., pH is a measure of the concentration of hydrogen ions in the water.

This measurement indicates the acidity or alkalinity of water. On the pH scale of 0 to 14, a reading of 7 is considered to be neutral. The value below 7 is recorded, as the water is more acidic while the value above 7 indicates thewater is alkaline. Normally, the value of pH ranging between 6 and 8 is considered as good water.

The solubility and availability of nutrients depend on upon the pH value, which is important for aquatic organisms.

2. Alkalinity

Alkalinity is due to the presence of bicarbonates, carbonates and hydroxides. Alkalinity is determined by thepotentiometric method or using apH meter of titrimetry using different acid-base indicators. Determination of water alkalinity in water is needed in water softening, chemical treatment of water.

3. Hardness

If the water does not produce lather with soap, it is called as *hard water*. The property is known as hardness. But it will produce a scummy white precipitate. The hardness is due to Ca^{2+}, Mg^{2+} and SO_4^{2-} , Cl^- , CO_3^{2-}, HCO_3^- ions and their salts.

4. Dissolved Oxygen

Oxygen is as important to life in water, as it is to live on land. Most aquatic plants and animals require for survival and the availability of oxygen affects their growth and development.

DO is the amount of oxygen dissolved in a given quality of water at a particular temperature and atmospheric pressure. DO is averyimportant to themeasure of the quality of water. The presence of oxygen in water is a positive sign and the absence oxygen is often a sign that the water is grossly polluted.

BOD (Biological Oxygen Demand)

BOD is one of the very important parameters to determine the quality of water. The sewage and wastewater from the industries often contain organic materials that are decomposed by microorganisms, which use the dissolved oxygen in the process. BOD is defined as "The amount of oxygen consumed by the microorganisms during the biochemical degradation of organic matter under aerobic conditions at 20°C for five days". BOD directly affects the amount of dissolved oxygen in rivers and streams. The greater the BOD, more will be the rapid depletion of oxygen in the stream. This means less oxygen is available for higher forms of aquatic life. BOD is expressed in mg/L or ppm.

Why is BOD important?

1. The consequences of high BOD are the same as those for dissolved oxygen. The higher the BOD, higher will be the stress, suffocation to the microorganism.
2. Water samples having a BOD value greater than 10 mg/L are considered to be highly polluted.

COD (Chemical Oxygen Demand)

COD is defined as the amount of oxygen required for the oxidation of organic matter as well as oxidizable inorganic matter. COD is expressed as mg/L.

Faecal Coliform Bacteria

Fecal Coliform Bacteria are microscopic organisms that live in the intestines of all warm-blooded animals and lives, animal wastes as faeces, eliminated from theintestinal tract. Faecal coliform bacteria may indicate the presence of disease-causing microorganisms, which live in the same environment as the Fecal Coliform bacteria.

The measurement is expressed as the number of organisms per 100 mL sample water. Drinking water should be free from pathogenic microorganisms.

Total Dissolved Solids

Total Dissolved Solids is a measure of the amount of particulate solids that are in solution which includes chlorides, nitrates, sulphates, phosphates of Na, K, Ca, Mg, Fe, Mn etc., A high concentration of total solids will make drinking water unpalatable and might have an adverse effect on people. Total solids also affect the clarity of thewater.

Total Dissolved Solids are measured by weighing the amount of solids present in a known volume of sample. This is done by gravimetric method. It is expressed in mg/L.

Nitrates

Nitrates are essential to plant nutrients, but in excess amounts, they can cause significant water quality problems.

Nitrates in combination with phosphate may accelerate eutrophication, which in turn will affect the DO, temperature and other indicators.

Nitrate is measured using the nitrate electrode method. This method is similar in function to dissolved oxygen matter. Nitrate is expressed in mg/L.

Phosphorous

Both phosphorous and nitrogen are essential nutrients for the plants and animals. But, the excess amount of phosphorous can accelerate the plant growth, algal blooms, which may cause no dissolved oxygen.

The phosphorous can be measured using electronic spectrophotometer or colorimeter. The unit of phosphorous is mg/L.

Hardness of Water

Hardness is the characteristic property present in water, which prevents lather with soap. The water sample, which gives a ready lather with soap, is called soft water. Water, which does not produce ready lather with soap, is known as hard water.

The salts of bicarbonates, chlorides and sulphates of calcium and magnesium are responsible for hardness. The reaction of hardness producing salts with the soap is as follows.

$$2C_{17}H_{35}COONa + CaCl_2 \longrightarrow (C_{17}H_{35}COO)_2Ca + 2NaCl$$

Hardness producing substance   (Calcium Stearate)

Hardness of the Water Can Classify Into Two Types

Temporary Hardness

The temporary hardness is due to the presence of bicarbonates of calcium and magnesium. The temporary hardness can be removed by

(i) Boiling the water.

(ii) Adding lime to the water.

On boiling, the temporary hardness producing salts are settling down as their insoluble carbonates and hydroxides.

$$Ca(HCO_3)_2 \xrightarrow{\text{Heat}} CaCO_3 \downarrow + H_2O + CO_2$$

$$Mg(HCO_3)_2 \xrightarrow{\text{Heat}} Mg(OH)_2 \downarrow + 2CO_2 \uparrow$$

Permanent Hardness

Permanent hardness or non-carbonate hardness is caused by the presence of chlorides and sulphates of calcium, magnesium, aluminium etc.

There are two methods to remove the permanent hardness.

(i) Lime soda process.

(ii) Zeolite process.

Units of Hardness

The units used to express hardness are:

1. **Parts per million(ppm)** is the number of parts by weight of calcium carbonate equivalent hardness per million (10^6) parts of water.

2. **Milligrams per litre (mg/L)** is the number of milligrams of calcium carbonate equivalent hardness present per litre of water.

$$1PPM = 1 \text{ mg/L}$$

1mg/L = one part of Calcium carbonate equivalent hardness in 10^6 parts of water (1 litre).

3. **Degree clerk(°C)** is the part of calcium carbonate equivalent hardness per gallon of water.

4. **The degree of French (°Fr)** is the part of calcium carbonate equivalent hardness per 10^5 parts of water.

$$1 \text{ ppm} = 1\text{mg/L.}$$

Hardness is always used to express as equivalents of calcium carbonate. The choice of calcium carbonate is due to its molecular weight 100 and equivalent weight 50 which is found to be convenient for the calculation purpose. The concentration of hardness producing salts can be calculated by using the following formula.

$$\text{Equivalent of calcium carbonate} = \frac{\text{Weight of hardness producing salts in mg} \times 50}{\text{Equivalent weight of salt}}$$

The same formula can be written as follows.

$$\text{Equivalent of calcium carbonate} = \frac{\text{Weight of hardness producing salts in mg} \times 100}{\text{Molecular weight of salt}}$$

Problems

1. Calculate the temporary and permanent hardness of a sample of water containing the dissolved salts in 1 litre of hard water as given below:

Calcium bicarbonate	=	64.8 mg
Calcium sulphate	=	13.6 mg
Magnesium chloride	=	50 mg
Magnesium bicarbonate	=	14.6 mg

Solution

Temporary hardness due to(i) $Ca(HCO_3)_2$

Equivalent of $CaCO_3$ hardness $= \dfrac{64.8 \times 100}{162}$ (Molecular weight of calcium bicarbonate = 162)

$= 40$ ppm

(ii) $Mg(HCO_3)_2$

Equivalent of $CaCO_3$ hardness $= \dfrac{14.6 \times 100}{146}$ (Molecular weight of Magnesium bicarbonate = 146)

$= 10$ ppm

Temporary hardness is due to $Ca(HCO_3)_2$ and $Mg(HCO_3)_2$.

$$\text{Hence the temporary hardness of the given water sample} = 40 + 10 = 50 \text{ ppm.}$$

Permanent hardness due to,

(i) $CaSO_4$

$$\text{Equivalent of } CaCO_3 \text{ hardness} = \frac{13.6 \times 100}{136} \quad \text{(Molecular weight of Calcium Sulphate} = 136)$$

$$= 10 \text{ ppm}$$

(ii) $MgCl_2$

$$\text{Equivalent of } CaCO_3 \text{ hardness} = \frac{50 \times 100}{95} \quad \text{(Molecular weight of Magnesium Chloride} = 95)$$

$$= 52.6 \text{ ppm}$$

Permanent hardness present in the given water sample $= 52.6 + 10 = 62.6$ ppm

2. Calculate the temporary and permanent hardness of water sample containing the following analysis

$$
\begin{aligned}
Mg(HCO_3)_2 &= 146 \text{ mg/L} \\
Ca(HCO_3)_2 &= 81 \text{ mg/L} \\
CaSO_4 &= 68 \text{ mg/L} \\
CaCl_2 &= 11.1 \text{ mg/L} \\
MgCl_2 &= 9.5 \text{ mg/L}
\end{aligned}
$$

Solution:

Temporary hardness due to,

(i) $Mg(HCO_3)_2$

$$\text{Equivalent of } CaCO_3 \text{ hardness} = \frac{146 \times 100}{146} \quad \text{(Molecular weight of Magnesium bicarbonate} = 146)$$

$$= 100 \text{ ppm}$$

(ii) $Ca(HCO_3)_2$

$$\text{Equivalent of } CaCO_3 \text{ hardness} = \frac{81 \times 100}{162} \quad \text{(Molecular weight of Calcium bicarbonate} = 162)$$

$$= 50 \text{ ppm}$$

Temporary hardness $= 100 + 50 = 150$ ppm

Permanent hardness due to,

(i) $CaSO_4$

    Equivalent of $CaCO_3$ hardness $= \dfrac{68 \times 100}{136}$ (Molecular weight of Calcium Sulphate = 136)

$= 50$ ppm

(ii) $CaCl_2$

    Equivalent of $CaCO_3$ hardness $= \dfrac{11.1 \times 100}{111}$ (Molecular weight of Calcium Chloride = 111)

$= 10$ ppm

(iii) $MgCl_2$

    Equivalent of $CaCO_3$ hardness $= \dfrac{9.5 \times 100}{95}$ (Molecular weight of Magnesium Chloride = 95)

$= 10$ ppm

Permanent hardness $= 50 + 10 + 10 = 70$ ppm

Total hardness $=$ Temporary hardness + Permanent hardness

$= 150 + 70 = 220$ ppm.

Determination of Hardness

It is essential to determine the hardness of water to use the water in boilers as well as in industries. This is done by the following methods.

EDTA Method

Since EDTA is insoluble in water, the soluble disodium salt of EDTA is used as the permanent complexing reagent with the Ca2+ and Mg2+ ions of the hard water. The structures are given below

HOOCH₂C   CH₂COOH

N—CH₂—CH₂—N

HOOCH₂C   CH₂COOH

EDTA - Ethylenediamine Tetraacetic acid.

HOOCH₂C   CH₂COONa

N—CH₂—CH₂—N

NaOOCH₂C   CH₂COOH

Disodium salt of EDTA

EDTA forms a very stable complex with water in the pH range 8-10. Before titration, the added Eriochrome black–T indicator forms an unstable complex of wine red colour. After the titration, the unstable complex forms astable complex with EDTA. Formation of theunstable complex by hard water with Eriochorme black–T indicator is given below.

$$\begin{bmatrix} Ca^{2+} \\ Mg^{2+} \end{bmatrix} + EBT \xrightarrow[\text{buffer}]{\text{pH} = 8\text{-}10} \begin{bmatrix} Ca \\ Mg \end{bmatrix} EBT$$
wine red coloured
weak unstable complex

After titration, the sodium salt forms astable complex in place of the unstable complex formed by Eriochrome black-T indicator. At the end point, the colour changes from wine red to steel blue. This is due to the following reaction.

$$\begin{bmatrix} Ca \\ Mg \end{bmatrix} EBT + EBT \longrightarrow \begin{bmatrix} Ca \\ Mg \end{bmatrix} EDTA + EBT$$
wine red coloured   (stable complex) (Steel blue)
weak unstable complex

Experiment

1. Preparation of the Indicator

Dissolve 0.5 gm of Eriochorme Black-T and about 100 gm of sodium chloride in 20ml of water.

2. Preparation of Buffer Solution

Dissolves 16.9 gm ammonium chloride (NH_4Cl) in 143 ml of concentrated ammonia solution and dilute the solution with distilled water to 250 ml.

3. Preparation of Standard Hard Water

Dissolve 1 gm of dry calcium carbonate in dilute HCl and evaporate the solution completely. Dissolve the residue in distilled water and make up the solution to 1 litreof distilled water.

4. Preparation of EDTA Solution

Dissolve 4gm of thepure sodium salt of EDTA crystals in 1 litre of distilled water.

5. Standardisation of EDTA Solution

Pipette out 50ml of standard hard water in a clean conical flask. Add 10-15ml of buffer solution and few drops of EBT indicator. Titrate this solution against EDTA taken in the burette. The end point is the colour change from wine red to steel blue. Note the end point. Repeat the titrations for the concordant value. Let the volume of EDTA consumed be V_1 ml.

6. Estimation of Total Hardness

Pipette out 50 ml of the hard water sample into a clean conical flask. Add 10 ml of buffer solution to maintain the pH 10. Add a few drops of EBT indicator and titrate the solution against standard EDTA taken in the burette. The end point is the change of colour from wine to steel blue. Repeat the titration for the concordant value. Let the volume of EDTA consumed be V_2ml.

7. Estimation of Permanent Hardness

Take 250 ml of the hard water sample in a beaker and evaporate it to 50 ml. Filter the solution into a 100 ml standard flask. Make up the solution again to 250 ml by adding distilled water. Pipette out 50ml of the made-up solution into a clean conical flask. Add 100 ml of buffer solution to maintain the pH 10. Add a few drops of EBT indicator and titrate the solution against standard EDTA taken in the burette. The end point is acolour change from wine red to steel blue. Let the volume of EDTA consumed be V_3 ml.

Calculations

Total Hardness

1 ml of standard hard water = 1mg of $CaCO_3$ equivalent hardness

I) V_1ml of EDTA is consumed by 50 ml standard hard water

or

50 ml of standard hard water consumes V_1 ml of EDTA

V_1 ml of EDTA $=$ 50 mg of $CaCO_3$ equivalent hardness

1 ml of EDTA $=$ 50 / V_1 mg of $CaCO_3$ equivalent hardness

II) V_2 ml of EDTA is consumed by 50 ml of sample hard water

We know,

1 ml of EDTA = 50/V_1 mg of $CaCO_3$ equivalent hardness

V_2 ml of EDTA = 50/V_1 x V_2 mg of $CaCO_3$ equivalent hardness

Similarly,

50 ml of sample hard water

contain $= 50/V_1$ x V_2 mg of $CaCO_3$ equivalent hardness

Therefore,

1ml of sample hard water contains = 50/V_1 x V_2/50mg of $CaCO_3$ equivalent hardness

1000ml of sample hard water contains = 50/V_1 x V_2/50 x 1000 mg of $CaCO_3$ equivalent

= V_2/V_1 x 1000 mg of $CaCO_3$ equivalent hardness

$$\boxed{\text{Total hardness} = V_2/V_1 \text{ x 1000 mg of } CaCO_3 \text{ equivalent hardness}}$$

Permanent Hardness:

50 ml of the sample hard water after removing temporary hard ness consumes V_3 ml of EDTA.

1 ml of EDTA $=$ 50/V_1 mg of $CaCO_3$ equivalent hardness

V_3 ml of EDTA $=$ 50/V_1 x V_3 mg $CaCO_3$ equivalent hardness

50 ml of sample water after boiling

contain $=$ 50/V_1 x V_3 mg of $CaCO_3$ equivalent hardness

1ml of sample hard water contains = 50/V_1 x V_3/50mg of $CaCO_3$ equivalent hardness

1000 ml of sample water $=$ 50/V_1 x V_3/50 x 1000 mg/L

$=$ V_3/V_1 x 1000 mg/L of $CaCO_3$ equivalent hardness

$$\boxed{\text{Permanent Hardness} = V_3/V_1 \text{ x 1000 mg/L}}$$

Temporary hardness:

$$
\begin{aligned}
\text{Temporary hardness} \;&=\; \text{Total hardness} - \text{Permanent hardness} \\
&=\; (V_2/V_1 \times 1000) - (V_3/V_1 \times 1000) \\
&=\; 1000 \times (V_2/V_1) - (V_3/V_1) \text{ ppm} \\
&=\; \frac{(V_2 - V_3)}{V_1} \times 1000 \text{ ppm}
\end{aligned}
$$

Purification of Water for Domestic Use

Potable water should be free from objectionable impurities. Sewage or industrial wastes should not contaminate it. It should be free from disease-causing bacteria's (pathogens). For removing various types of impurities the following treatment processes are employed.

Screening ⟶ Aeration ⟶ Sedimentation ⟶ Coagulation

Sterilization ⟵ Filtration

1. Screening

The raw water, which contains suspended impurities, is passed through screens, having a large number of pores, when floating matters are retained by them.

2. Aeration

The process of mixing water with air is known as aeration. The main purpose of aeration is

(i) To remove gases like CO_2, H_2S and other volatile impurities causing bad taste and odour to water.

(ii) To remove ferrous and manganous salts as insoluble ferric and manganic salts.

3. Sedimentation

Sedimentation is the process of removing the sludge and other solid impurities by allowing the water to stay undisturbed for 2-6 hours. The clear supernatant water is then drawn from thetank with the help of pumps. The different types of sedimentation tanks are horizontal flow tank, radical flow circular tank and vertical flow hopper tank.

4. Coagulation

When water contains colloidal particles, it becomes necessary to apply sedimentation with coagulation to remove colloidal impurities. Coagulation is the process of removing colloidal matter from water with the addition of arequisite amount of coagulants. The chemical substances like alum [$K_2SO_4Al_2(SO_4)_3.24H_2O$], Aluminiumsulphate, [$Al_2(SO_4)_3.8H_2O$], ferrous sulphate [FeSO$_4$.7H$_2$O] etc are used as coagulants. These chemical substances react with carbonate and a bicarbonate ion present in water and form a precipitate which settles down.

$$Al_2(SO_4)_3 + 3Ca(HCO_3)_2 \longrightarrow 2Al(OH)_3 + 3CaSO_4 + 6CO_2$$

$$FeSO_4 + Mg(HCO_3)_2 \longrightarrow Fe(OH)_2 + MgCO_3 + CO_2 + H_2O$$

$$4Fe(OH)_2 + O_2 + 2H_2O \longrightarrow 4Fe(OH)_3$$

Fe(OH)$_3$ which is in the form of heavy flow causes quick sedimentation.

5. Filtration

Filtration is a process of removing colloidal and bacterial impurities by passing water through sand filters. The sand filters consist of athick layer of fine sand placed over several beds of coarse sand and gravels and water is allowed to percolate through the bed.

a) Sand Filter

The sand filter is made up of therectangular concrete tank in, which placed three feet of fine sand at the top, one foot of coarse sand in the middle and 8 inches thick of, graded gravel at the bottom as the filtering material. Sediment-water entering the sand filter is uniformly distributed over theentire sand bed. When water percolates through the filter most of the objectionable impurities are retained by the sand bed. When the rate of filtration becomes slow the fine sand bed has to be removed, washed and again replaced. The sand filter method is an effective efficient process of municipal water supply.

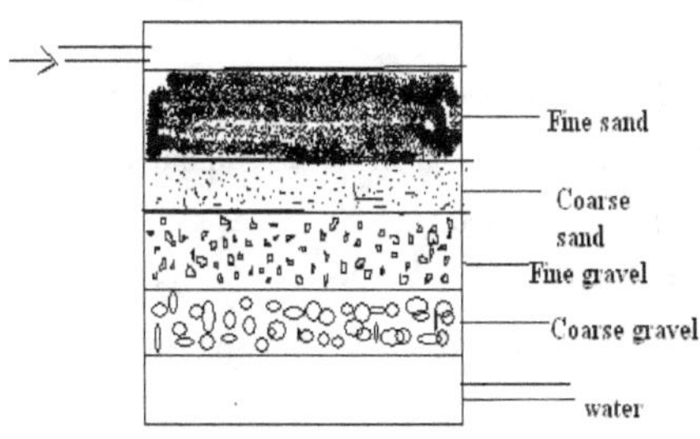

Sand Filter

6. Sterilisation of Water

The process of destroying the disease-producing bacteria's and microorganisms from the water and making it as portable water is called disinfection. There are various methods of disinfection of water.

1. **By boiling** water for 20 minutes all the disease producing bacteria are killed and water becomes safe for use. But, boiling alters the taste of drinking water and also it is impossible to employ it in municipal water levels.

2. Sterilisation of water can be done by using ultraviolet rays. In this process ultra, violetrays are concentrated on flowing water. The UV radiation kills bacteria's. But it is costlier method when applied to large quantities of water.

Using ozone: Ozone is a powerful disinfectant and is readily absorbed by water.

Ozone is highly unstable and breaks down to give nascent oxygen.

$$O_3 \longrightarrow O_2 + [O]$$
$$\text{Nascent oxygen}$$

The nascent oxygen is a powerful oxidising agent and kills the bacteria.

Disadvantages

1. This process is costly and cannot be used in large scale.
2. Ozone is unstable and cannot be stored for a long time.

Chlorination

The process of adding chlorine to water is called chlorination. Chlorination can be done by the following methods.

i. **By adding Chlorine gas:** Chlorine gas can be bubbled through the water as a very good disinfectant.

$$Cl_2 + H_2O \longrightarrow HCl + HOCl \text{ (Hypochlrous acid)}$$

ii. **By adding Chloramine:** When chlorine and ammonia are mixed in the ratio 2:1 compound chloramines is formed.

$$Cl_2 + NH_3 \longrightarrow ClNH_2 + HCl$$

iii. Chloramine compounds decompose slowly to give chlorine. It is a better disinfection than chlorine.

By Adding Bleaching Powder

By adding bleaching powder (Calcium hypochlorite) the disease-causing bacteria can also be killed. The added bleaching powder reacts with water to produce hypochlorous acid, a powerful disinfectant.

$$CaOCl_2 + H_2O \longrightarrow Ca(OH)_2 + Cl_2$$

$$Cl_2 + H_2O \longrightarrow HCl + HOCl \text{ (Hypochlrous acid)}$$

$$Pathogens + HOCl \longrightarrow Pathogens \text{ are killed}$$

The disinfecting action of bleaching power is due to the chlorine produced when it reacts with water. Chlorination is the most effective method used for municipal water supply.

Break Point Chlorination

Breakpoint chlorination is the application of chlorine to produce a residual free available chlorine with no combined chlorine present. As chlorine is added to water, it reacts with ammonia and forms chloramines. Hence, chlorine in the combined form gradually increases. After a particular limit oxidation of chloroamine and other impurities start and there is a fall in the combined chlorine state. When all combined chlorine has been oxidised by thereaction, with free available chlorine, the residual, now consisting of only of free available chlorine rises again and continues to increase in direct proportion to increased dosage. The point at which the residual again begins to increase is the break point.

The breakpoint chlorination curve appears in four stages. Stage 'a' shows a typical breakpoint for water containing a considerable amount of ammonia. During the initial upward rise, chloramines are first formed. The curve rises until sufficient free available chlorine is developed to react with chloramines, then it falls until a point where all ammonia compounds have been oxidised.

While less organic matter in the water, as in stage 'b' and 'c' free available chlorine is formed sooner, destroying chloramines formed at the early stage.This results in lower combined chlorine residuals and flatters curves before breakpoint. The stage' shows the chloramines are neutralised at an early stage by the upswing of the curve.

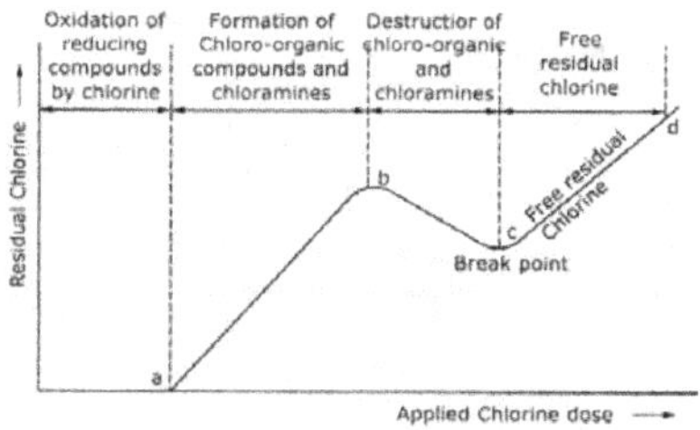

Break point chlorination. (Graph)

Advantages of Break Point Chlorination

1. Highly bactericidal efficiency.
2. It removes colour and odour from thewater.
3. It oxidises completely the organic compounds present in the water.

1.16. Demineralisation Processor ION Exchange Process

This process removes all the ions (cations & anions)present in the hard water. Soft water produced by lime-soda process contains Na^+, K^+, SO_4^{2-}, Cl^- ions, but don't have the salts of Ca^{2+} / Mg^{2+} ions. In another term DM water does not contain both the ions. *Thus, soft water isnot DM water where as DM is soft water.* This process is carried out by using ion exchange resins which are along chain, cross-linked, insoluble organic polymers with micro-porous structure. The functional group attached to this chains are responsible for Ion exchanging properties.

1. Cation Exchanger

Resins containing acidic functional groups(-COOH, -SO_3H) are capable of exchanging their H^+ ions with other cations of hard water. Cation exchange resin isrepresented as RH2 .

1. *Sulphonated coals.*
2. *Sulphonated polystyrene.*

$$R\text{-}SO_3H; R\text{-}COOH \equiv RH2$$

2. Anion Exchanger

Resins containing basic functional groups (-NH_2-OH) are capable of exchanging their anions with other anions of hard water. Anion exchange resin is represented as R $(OH)_2$

1. *Cross-linked quaternary ammonium salts.*
2. *Urea -formaldehyde resin.*

$$R\text{ - }NR_3OH \; ; \; R\text{-}OH \; ; \; \mathbf{R\text{-}NH_2} \equiv R(OH)_2$$

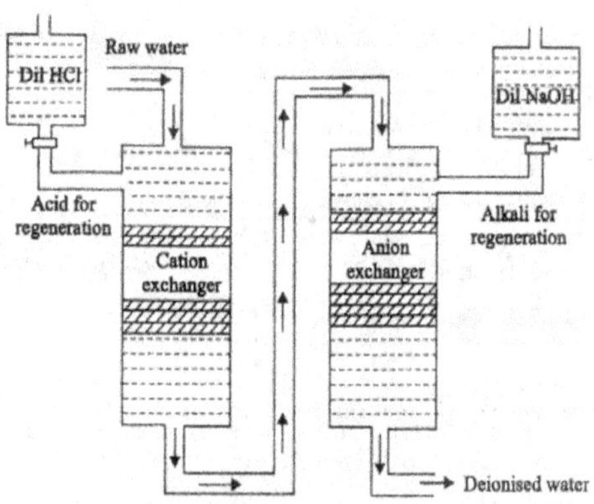

Demineralisation Process

Process

The hard water first passed through a cation exchange column, (Fig. 1.5) which absorbs all the cations likeCa^{2+}, Mg^{2+}, Na$^+$, K$^+$, etc., present in the hard water.

$$RH_2 + CaCl_2 \longrightarrow RCa + 2HCl$$

$$RH_2 + CaSO_4 \longrightarrow RCa + H_2SO_4$$

$$RH_2 + MgCl_2 \longrightarrow RMg + 2HCl$$

$$RH_2 + MgSO_4 \longrightarrow RMg + H_2SO_4$$

$$RH + NaCl \longrightarrow RNa + HCl$$

The cation-free water is then passed through anions exchange column, which absorbs all the anions like Cl$^-$, SO$_4{}^{2-}$, etc., present in the water.

$$R'(OH)_2 + HCl \longrightarrow R'Cl_2 + 2H_2O$$

$$R'(OH)_2 + H_2SO_4 \longrightarrow R'SO_4 + 2H_2O$$

The water coming out of the anion exchanger iscompletely free from cations and anions. This water is knownasdemineralized water or deionised water.

Regeneration

When the cation exchange resin is exhausted, it can be Sealregenerated by passing a solution of dilHCl or dil H$_2$SO$_4$

$$R'Ca + 2HCl \longrightarrow R'H_2 + CaCl_2$$

$$R'Na + 2HCl \longrightarrow R'H_2 + NaCl$$

Similarly, when the anion exchange resin is exhausted.it can be regenerated by passing a solution of NaOH.

$$R'Cl_2 + 2NaOH \longrightarrow R'(OH)_2 + 2NaCl$$

Advantages of Ion-Exchange Process

1. Highly acidic or alkaline water can be treated by this process.
2. The water obtained by this process will have avery low hardness (nearly 2 ppm).

Disadvantages of Ion-Exchange Process

1. Water containing turbidity, Fe and Mn cannot be treated, because turbidity reduces the output and Fe, Mn forms astable compound with the resin.
2. The equipment is costly and more expensive chemicals are needed.

Internal Conditioning

It involves the removal of scale forming substance, which was not completely removed in the external treatment, by adding chemicals directly into the boiler. These chemicals are also called boiler compounds.

1. Carbonate Conditioning

Scale formation can be avoided by adding Na_2CO_3 to the boiler water. It is used only in low-pressure boilers. The scale-forming salt like $CaSO_4$ is converted into $CaCO_3$, which can be removed easily.

$$CaSO_4 + Na_2CO_3 \longrightarrow CaCO_3 + Na_2SO_4.$$

Desalination of Brackish Water

1. The process of removing the **common salt** from the water is known as desalination. The water containing dissolved salts (high percentage of NaCl) with a salty taste is called brackish water eg., sea water.

2. In oil-rich west Asian countries, which are surrounded by sea, contain no potable water but only seawater, and brackish water. Brackish water is totally unfit for drinking purpose. But, the sea water can be made available as drinking water through desalination process. Commonly used methods for the desalination of brackish water are reverse osmosis and electrodialysis.

Reverse Osmosis

When a semi-permeable membrane separates two solutions of different concentrations, theflow of solvent takes place from dilute to concentrated solution due to osmosis. But pressure in excess of osmotic pressure has applied the water to flow in the reverse direction to the direction of flow in natural osmosis. This process is called reverse osmosis. This process is also known as hyperfiltration.

Reverse osmosis-from a concentrated solution of a dilute solution. Reverse osmosis is used in conjunction with one or more activated carbon filters; this combination removes the 99.9 % of undesirable water contaminants.

In this process high pressure of the water 15-40 kg/cm² applied to the raw water to force the water to pass through a thin film reverse osmosis membrane, leaving behind the dissolved solids. Some of the thin film membranes used in reverse osmosis process are cellulose acetate, polysulphone, polyamide etc. In the latest technique, hydrophobic membranes are used as asemi-permeable membrane.

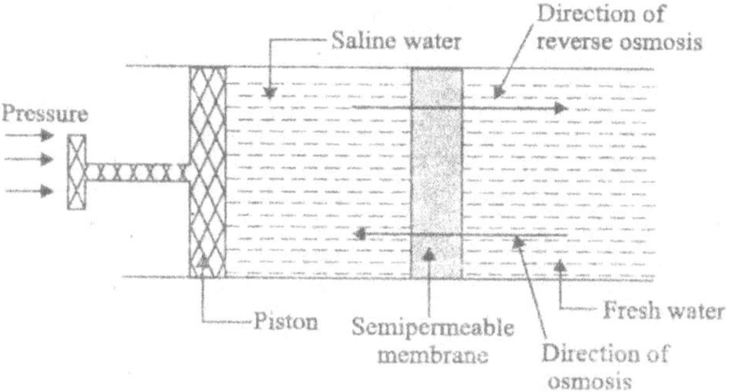

Advantages of Reverse Osmosis

1. High lifetime.
2. Removes ionic, non-ionic and colloidal silica impurities, which can not be removed by demineralization method.
3. Low capital cost.
4. Simple operational procedure.
5. The membrane can be replaced within a few minutes, thereby providing uninterrupted water supply.

UNIT II

CHEMISTRY FOR ENGINEERING PLASTICS

Polymers play an important role in human life. In fact, our body is made of a lot of polymers (e.g.) proteins, enzyme etc.

Polymers are used in the field of automobiles, Electrical & electronics, telecom, railways etc. The polymer like polybutylene terephthalate is used in electrical and electronics

Engineers often use the term "polymer" when referring to theplastic material, because it more clearly describes how many (poly) chemical units (mers) form up in thecomplex chain to create modern plastic resins.

Most of the people think plastic means cheap and breakable, but when engineers search for new ways to enhance weight savings corrosion resistance, shock and vibration dampening and stealth they immediately turn to plastic.

Therefore, engineers working in any field need to understand the concepts of polymer process and also polymer product performance.

The word polymer is derived from Greek words "poly" means many and "mers" means units or parts. Polymers are macromolecules or giant molecules built-up by linking together of a large number of small molecules.

Monomer: The smallest repeating unit of a polymer. The monomer is amicro molecule which combines with each other to form a polymer.

Polymer: Polymer ismacromolecules (giant molecules of higher molecular weight) formed by the repeated linking of a large number of small molecules called monomer.

For example, Polyethylene is a polymer formed by linking together of a large number of ethane molecules by means of acovalent bond.

Examples of monomer and polymers.

Nomenclature of Polymer

- Homopolymer.
- Heteropolymer (or) copolymer.
- Homochain polymer.
- Heterochain polymer.

Monomer	Polymer
ethylene	polyethylene
propylene	polypropylene
butene-1	polybutene
vinyl chloride	polyvinyl chloride

Homopolymer

A polymer containing thesame type of monomers is known as homopolymers.

Example:polyethene

$$-CH_2-CH_2-CH_2-CH_2-CH_2-CH_2-$$

Copolymer (or) Heteropolymer

A polymer containing more than one type of monomers is known as copolymers.

Example : Nylon 6:6.

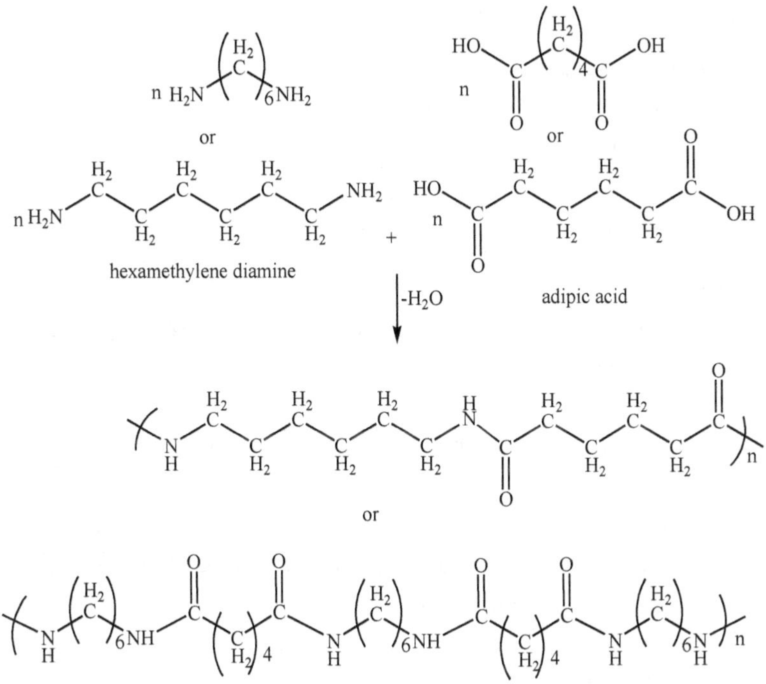

hexamethylene diamine

adipic acid

$-H_2O$

or

Homo Chain Polymer

- If the main chain is made is up of the same species of atoms the polymers are called homochain polymer.

Example polyvinyl chloride.

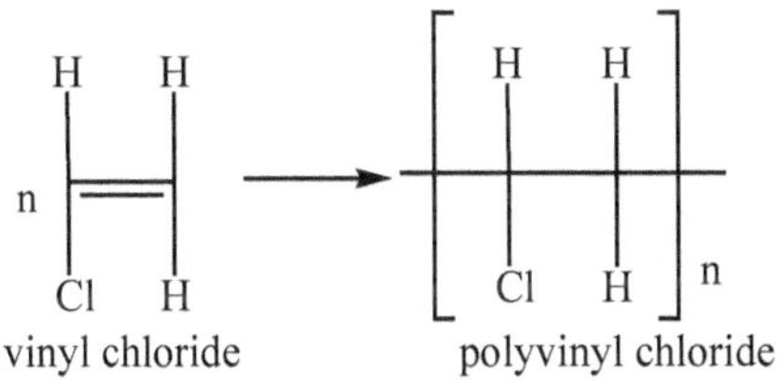

vinyl chloride   polyvinyl chloride

Heterochain Polymer

- If the main chain of polymers is made up of different atoms it is called heterochain polymer.

Example: -C-C-O-C-C-O-C-C-O-C-C-main chain.

Tacticity

The orientation of monomeric units in a polymer molecule can take place in an orderly or disorderly fashion with respect to the main chain. The difference in configuration (tacticity) does affect their properties.

- Isotactic polymer.
- Atactic polymer and Syndiotactic polymer.

Functionality

- The number of bonding sites or reactive sites or functional groups present in a monomer is known as its functionality

Example	Functionality
ethylene	2 (one double bond is present, hence bifunctional monomer)
Hexamethylene diamine	2 (2 functional groups are present, hence bifunctional monomer)
glycerol	3 (3 functional groups are present, hence trifunctional monomer)

Significance

- Bi functional monomer.
- Tri functional monomer.
- Poly functional monomer.

Bifunctional Monomers

- The functionality of bifunctional monomer is two.
- Mainly forms linear (or) straight chain polymer.
- Each monomeric unit in the linear chain is linked by thestrong covalent bond (primary bonds).

- But the different chains are held together by a weak Van der Waals force of attraction (secondary bonds)

Bi functional monomer

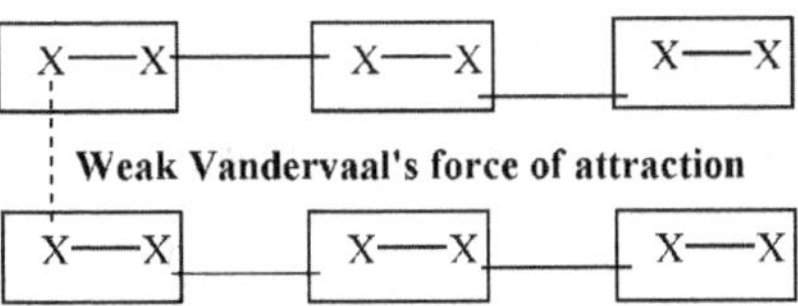

Trifunctional Monomers

- The functionality of trifunctional monomer is three.
- When a trifunctional monomer is mixed in asmall amount with a bifunctional monomer. They form branched chain polymer.

Tri functional

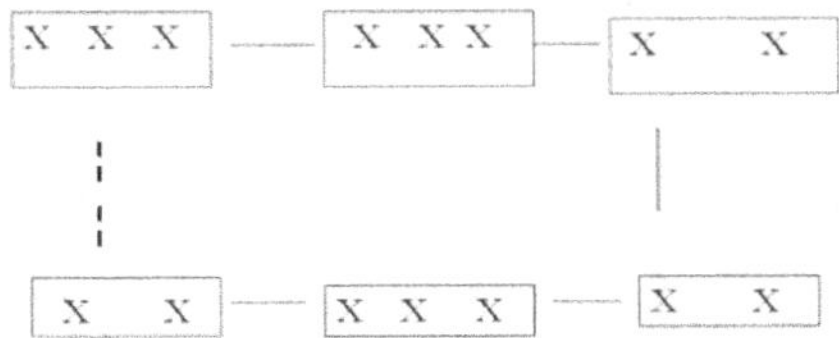

Polyfunctional Monomers

- Functionality of polyfunctional monomer depends on the nature of monomer
- It forms cross-linked polymer three- dimensional network polymer. All the monomers in the polymer are connected to each other by strong covalent bonds.

Poly functional

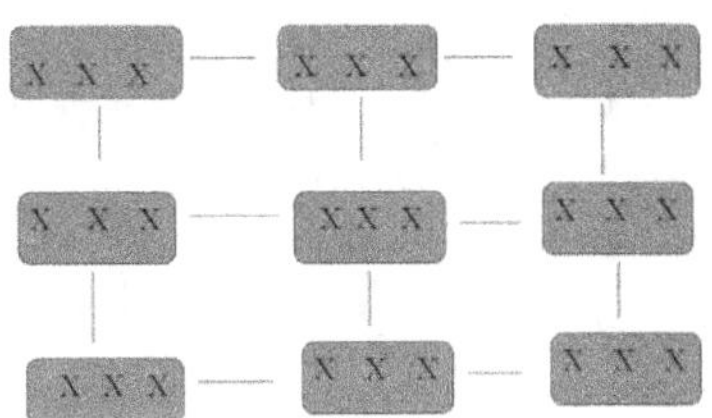

Polymerization

Polymerization is a chemical reaction in which a large number of smaller molecules combine to give a macromolecules with or without elimination of small molecules like water, HCl.

Degree of Polymerization, DP

Thenumber of repeating units (n) or monomer present in the polymer chain is called the degree of polymerization.

The degree of polymerization is the ratio between themolecular weight of the polymeric network and molecular weight ofthe repeating unit.

$$DP = \text{average number of repeat units per chain}$$

$$DP = \frac{\overline{M}_n}{m}$$

M_n - molecular weight of the polymeric network.

m- Molecular weight of the repeating unit.

$$2\ CH_2{=}CH_2 \longrightarrow {-}(CH_2{-}CH_2){-}\ (CH_2{-}CH_2){-}$$

In this example 2, repeating units are present in the polymer chain . So the degree of polymerization is 2.

Oligo polymer – the polymer with low degree of polymerization are known as oligopolymer their molecular weight ranges from 500-5000 .

High polymer- the polymer with a high degree of polymerization are known as high polymers and their molecular weight ranges from   10,000-2, 00,000

Types of Polymerization

- **Addition polymerization or chain growth polymerization.**

  It is the reaction that yields a polymer, which is an exact multiple of the original monomeric molecule. The original monomeric molecule, usually contains C=C bonds Double bond opens to (link) bond to next monomer molecule Chain forms when thesame basic unit is repeated over and over. There is no elimination of any molecules.

- **Example: 1 Tetrafluoroethene** can be polymerised to form polytetrafluoroethylene, commonly known as Teflon.

$$CF_2=CF_2 \longrightarrow \cdots \left[\begin{array}{c} | \\ -C-C-C-C- \\ | \end{array} \right]_n \cdots$$

- **Example 2 PVC**

$$CH_2=CHCl \longrightarrow \left[-CH(Cl)-CH_2- \right]_n$$

Name(s)	Formula	Monomer	Properties	Uses
Polyethylene low density (LDPE)	$-(CH_2\text{-}CH_2)_n-$	ethylene $CH_2=CH_2$	soft, waxy solid	film wrap, plastic bags
Polyethylene high density (HDPE)	$-(CH_2\text{-}CH_2)_n-$	ethylene $CH_2=CH_2$	rigid, translucent solid	electrical insulation bottles, toys
Polypropylene (PP) different grades	$-[CH_2\text{-}CH(CH_3)]_n-$	propylene $CH_2=CHCH_3$	<u>atactic</u>: soft, elastic solid <u>isotactic</u>: hard, strong solid	similar to LDPE carpet, upholstery
Poly(vinyl chloride) (PVC)	$-(CH_2\text{-}CHCl)_n-$	vinyl chloride $CH_2=CHCl$	strong rigid solid	pipes, siding, flooring
Poly(vinylidene chloride) (Saran A)	$-(CH_2\text{-}CCl_2)_n-$	vinylidene chloride $CH_2=CCl_2$	dense, high-melting solid	seat covers, films
Polystyrene (PS)	$-[CH_2\text{-}CH(C_6H_5)]_n-$	styrene $CH_2=CHC_6H_5$	hard, rigid, clear solid soluble in organic solvents	toys, cabinets packaging (foamed)
Polyacrylonitrile (PAN, Orlon, Acrilan)	$-(CH_2\text{-}CHCN)_n-$	acrylonitrile $CH_2=CHCN$	high-melting solid soluble in organic solvents	rugs, blankets clothing

Name(s)	Formula	Monomer	Properties	Uses
Polytetrafluoroethylene (PTFE, Teflon)	$-(CF_2\text{-}CF_2)_n-$	tetrafluoroethylene $CF_2=CF_2$	resistant, smooth solid	non-stick surfaces electrical insulation
Poly(methyl methacrylate) (PMMA, Lucite, Plexiglas)	$-[CH_2\text{-}C(CH_3)CO_2CH_3]_n-$	methyl methacrylate $CH_2=C(CH_3)CO_2CH_3$	hard, transparent solid	lighting covers, signs skylights
Poly(vinyl acetate) (PVAc)	$-(CH_2\text{-}CHOCOCH_3)_n-$	vinyl acetate $CH_2=CHOCOCH_3$	soft, sticky solid	latex paints, adhesives
cis-Polyisoprene natural rubber	$-[CH_2\text{-}CH=C(CH_3)\text{-}CH_2]_n-$	isoprene $CH_2=CH\text{-}C(CH_3)=CH_2$	soft, sticky solid	requires vulcanization for practical use
Polychloroprene (cis + trans) (Neoprene)	$-[CH_2\text{-}CH=CCl\text{-}CH_2]_n-$	chloroprene $CH_2=CH\text{-}CCl=CH_2$	tough, rubbery solid	synthetic rubber oil resistant

Condensation or Step-Wise Polymerization

It is a reaction between simple polar groups containing monomers with the formation of polymer and elimination of small molecules like $H_2O.HCl$ In order to form a condensation polymer the monomer must have two functional groups.

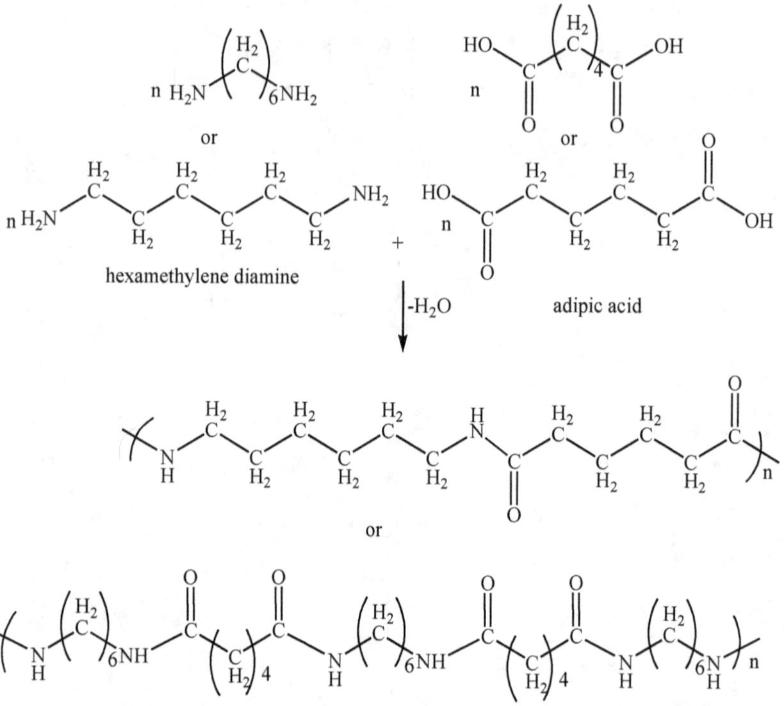

hexamethylene diamine

adipic acid

$-H_2O$

In some cases, condensation polymerization takes place without the elimination of small molecules like H_2O, HCl, etc but by just the opening of cyclic compounds.

Example 2 Polymerization of Caprolactam to form Nylon 6 Co-Polymerization

It is the joint polymerization in which two or more different monomers combine to give a polymer.

High molecular weight polymers obtained by copolymerization are called copolymers. Copolymerization is mainly carried out to vary the properties of polymers such as hardness, strength, rigidity, heat resistance etc.

Example butadiene and styrene copolymerize to give GR-S rubber

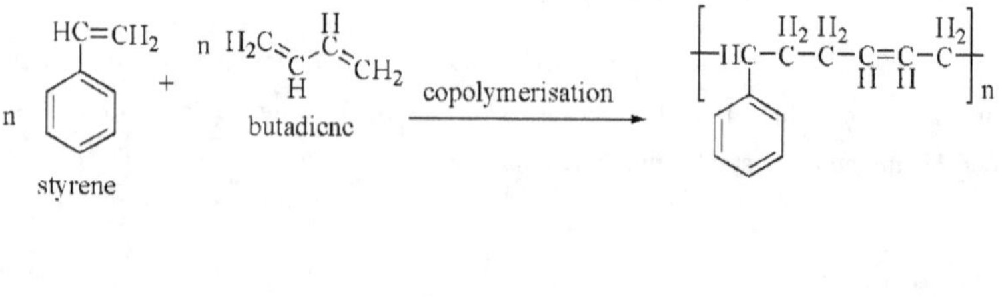

styrene

butadiene

copolymerisation

Difference between Addition and Condensation Polymerization

Addition polymerization	Condensation polymerization
The monomer must have at least one multiple bonds. **Example :** Ethylene $CH_2=CH_2$ Acetylene $CH\equiv CH$	The monomer must have at least two identical (or) different functional groups. **Example** glycol $\begin{array}{c} CH_2\text{-}OH \\ | \\ CH_2\text{-}OH \end{array}$ **6-amino hexanoic acid** $H_2N\text{-}(CH_2)_6\text{-}COOH$
Monomers add on to give a polymer and no other by-product is formed.	Monomers combine togive a polymer and by-product such as H_2O, CH_3OH are formed.
The number of monomeric units decreases steadily throughout the reaction.	Monomers disappear at the early stage of thereaction.
Homo chain polymer is obtained	Heterochain polymer is obtained
The molecular weight of the polymer is an integral multiple of molecular weight of monomer.	The molecular weight of the polymer need not be an integral multiple of monomer.
Addition polymerization	**Condensation polymerization**
Thermoplastics are produced e.g polyethene ,PVC	Thermosetting plastics are produced. (E.g) Bakelite, urea formaldehyde.
High molecular weight polymer is formed at once.	The molecular weight of the polymer rises steadily throughout the reaction.
Longer reaction times give higher yield but, have a little effect on molecular weight.	Longer reaction times are essential to obtain high molecular weight.

Mechanism of Addition Polymerization

The mechanism of addition polymerization can be explained by any one of the following three types.

1. Free radical mechanism.

2. Ionic mechanism.

3. Coordination mechanism.

Free radical mechanism occurs in three major steps namely,

1. Initiation.

2. Propagation.

3. Termination.

Initiators are compounds which produce free radicals by the homolytic dissociation. These free radicals initiate the polymer chain growth.

Initiators: Initiators are compounds which produce free radicals by the homolytic dissociation there free radicals initiate the polymer chain growth.

Initiation

It isconsidered to involve two reactions,

(a)   The first reaction involves production of free radicals by homolytic dissociation of an initiator (or) catalyst to yield a pair of free radicals (R)∘

$$I \longrightarrow 2R∘$$

Initiator   free radical

Some commonly used thermal initiators are benzoyl peroxide and acetyl peroxide.

Thermal initiators are a substance used to produce free radicals by homolytic dissociation at high temperature. (70-90⁰C)

$$CH_3COO\text{-}OOCCH_3 \longrightarrow 2CH_3COO^{\bullet} \text{ (or) } 2R^{\bullet}$$
$$C_6H_5COO\text{-}OOCC_6H_5 \longrightarrow 2C_6H_5COO^{\bullet} \text{ (or) } 2R^{\bullet}$$

Example : Free radical mechanism of ethylene

Step 1: Initiation

$$R\text{---}R \longrightarrow 2R^{\bullet}$$

$$R^{\bullet} + H_2C{=}CH_2 \longrightarrow R\text{---}\overset{H_2}{C}\text{---}\overset{\bullet}{C}H_2$$

Step: 2 Propagation

It involves the growth of chain initiating species by successive addition of a large number of monomers. The growing chain of the polymer is known as living polymer.

$$R\text{---}\overset{H_2}{C}\text{---}\overset{\bullet}{C}H_2 + H_2C{=}CH_2 \longrightarrow R\text{---}\overset{H_2}{C}\text{---}CH_2\text{---}\overset{H_2}{C}\text{---}\overset{\bullet}{C}H_2$$

$$R\text{---}\overset{H_2}{C}\text{---}CH_2\text{---}\overset{H_2}{C}\text{---}\overset{\bullet}{C}H_2 + H_2C{=}CH_2 \longrightarrow R\text{---}\overset{H_2}{C}\text{---}CH_2\text{---}\overset{H_2}{C}\text{---}CH_2\text{---}\overset{H_2}{C}\text{---}\overset{\bullet}{C}H_2$$

Step: 3

Termination of the growing chain of thepolymer may occur either by coupling or disproportionation.

a) *Coupling (or) Combination*

It involves thecoupling of free radical of one chain end to another free radical forming a macromolecule.

b) *Disproportionation*

It involves thetransfer of a hydrogen atom of one radical centre to another centre forming two macromolecules one saturated and another unsaturated.

Coupling reaction

$$R{\left(\overset{H_2}{\underset{}{C}}-CH_2\right)}_n\overset{H_2}{C}-CH_2^{\bullet} \; + \; {}^{\bullet}H_2C-CH_2{\left(\overset{H_2}{C}-\overset{H_2}{C}\right)}_n R$$

$$\downarrow$$

$$R{\left(\overset{H_2}{C}-CH_2\right)}_n\overset{H_2}{C}-CH_2-\overset{H_2}{C}-CH_2{\left(\overset{H_2}{C}-\overset{H_2}{C}\right)}_n R$$

or

Disproportionation reaction

$$R{\left(\overset{H_2}{C}-CH_2\right)}_n\overset{H_2}{C}-\overset{\bullet}{C}H_2 \; + \; \overset{H}{\underset{H}{C}}-CH_2{\left(\overset{H_2}{C}-\overset{H_2}{C}\right)}_n R$$

$$\downarrow$$

$$R{\left(\overset{H_2}{C}-CH_2\right)}_n\overset{H_2}{C}-CH_3 \; + \; H_2C{=}CH-{\left(\overset{H_2}{C}-\overset{H_2}{C}\right)}_n R$$

The product of addition polymerization is known as adead polymer.

2.1. Plastics

The term plastics or plastics material, in general, is given to organic materials of high molecular weight, which can be moulded into any desired shape by the application heat and pressure in the presence of a catalyst.

The term plastic must be differentiated from the resin. The resin is the basic binding materials, which form a major part of the plastics, and which actually has undergone polymerization and condensation reactions.

Advantages

1. Lightweight.
2. Low melting point.
3. Corrosion resistance.
4. Easily moulded.
5. Impermeable to water.

Disadvantages

1. Very high softness.
2. Embrittlement at low temperature.
3. Deformation under load.
4. Easily combustible.
5. Low heat resistance.

Uses

- For making electrical goods.
- In aeronautical engineering.
- For making furniture.
- For making electrical appliances such as plugs , switches, holders and T.V cabinets.

Classification of plastics

Based on structure	Based on usage
Thermoplastics eg. Polyethylene	General purpose plastics eg. Polyethene, polypropylene
Thermosetting plastics Eg. Bakelite, polyester	Engineering plastics Eg. PVC, Teflon

2.2. Based on Structure

Thermoplastic Resin

This type of plastics is prepared by addition polymerization. They are straight chain (or) slightly branched polymers and various chains are held together by weak Vander Waals force of attraction. Thermoplastic can be softened by heating and hardened on cooling. They are generally soluble in organic solvents.

Example; PVC

$$-M-\ M-\ M-\ M-\ M-\ M\ -$$

$$-M-\ M-\ M-\ M-\ M-\ M-$$

Thermosetting Resins

Thermosetting plastics are prepared by condensation polymerization. various chains are held together by strong covalent bonds.

Thermosetting plastics get harden on heating and once harden they cannot be softened again. They are almost insoluble in organic solvents.

Example; Bakelite, polyester.

$$-M- M- M- M- M- M -$$
$$|\qquad|\qquad|$$
$$-M- M- M- M- M- M-$$

Difference between Thermoplastics and Thermosetting Plastics

Thermoplastics	Thermosetting plastics
They are formed by addition polymerization	They are formed by condensation polymerization
They consist of linear long chain polymers	They consist of three-dimensional network structure
All the polymer chains are held together by Weak Vander walls forces.	All the polymer chains are linked by strong covalent bonds.
They are weak soft and less brittle	They are strong, hard and more brittle
They soften on heating and harden on cooling	They do not soften on heating
They can be remoulded	They cannot be remoulded
They have low molecular weights	They have high molecular weights
They have soluble in organic solvents	They are insoluble in organic solvents

2.3.   Classification based on Usage

General Purpose Plastics

General purpose plastics have low to medium mechanical properties . They are used for the manufacture of commodity items. They account for about 80-85% of polymer production.

Example

Polyethene, polypropylene, PVC

Properties of General Purpose Plastics

- They are mostly crystalline with low glass transition temperature (or) they are glossy (or) amorphous polymer.
- They have low use temperature; therefore, they cannot be used at high temperature.
- They have low to medium mechanical properties.
- Generally, they have low abrasion resistance and poor dimensional stability.

Ex : polypropylene, polyvinyl chloride.

Engineering Plastics

(i) They have high mechanical strength, abrasion resistance.

(ii) High load bearing capacity.

(iii) High dimensional stability.

(iv) Readily moldable properties.

(v) Can't broken easily.

(vi) Ex : PVC .Nylon 6,6 ,Teflon.

Applications

(i) They can be used alone or in conjunction with metals, ceramics or glasses etc.

(ii) They find applications in demanding areas like automobiles, defence, electrical, telecommunications, textiles, satellite, robots, computer components etc.

2.4.   Important Engineering Plastics

Polyvinyl Chloride (PVC) Preparation

Preparation of PVC involves the following two steps.

Step:1 Preparation of Vinyl Chloride

Vinyl chloride is prepared by treating acetylene with hydrogen chloride at 60-80⁰C in the presence of metal chloride as acatalyst.

$$HC\equiv CH + HCl \longrightarrow \underset{}{CH_2=CHCl}$$

acetylene

Step:2 Preparation of PVC from Vinyl Chloride

Polyvinyl chloride is obtained by heating water emulsion of vinyl chloride in thepresence of benzoyl peroxide (or) hydrogen peroxide under pressure.

$$n\ CH_2=CHCl \xrightarrow{polymerisation} -(CH_2-CHCl)_n-$$

Properties

(i) PVC is a colourless, odourless and chemically inert powder.

(ii) It is soluble in inorganic acids and alkalis but soluble in hot chlorinated solvents.

(iii) It undergoes degradation in presence of sunlight.

Uses

(i)   It is used in the production of pipes cables insulations, table covers and rain coats etc.

(ii)   It is used also used for making sheets which are employed for tank linings, light fittings, refrigerator components etc.

Teflon (or) Fluon (or) Polytetrafluoroethylene (PTFE) Preparation

Polymerization of water emulsion of tetrafluoroethylenegives   Teflon in Presence of benzoyl peroxide $(C_6H_5CO_2)_2O_2$

Properties

(i)   Since the fluorine atoms are the strong electronegative elements they tightly bond with carbon atoms in Teflon .As (C-F) bond is stronger it is non-reactive is not wetted by oil and H_2O. so Teflon is non-sticky.

(ii)   It is extremely tough flexible material possessing high softening point.

(iii)   It possesses good electrical and mechanical properties.

(iv)   Chemically resistant towards all chemicals.

(v)   Excellent thermal stability low co-efficient of friction.

Uses

(i)   Used as good electrical insulating materials in motors, cables, transformers.Used in making gaskets, pump parts, tank linings.

(ii)   It is also used in chemical carrying pipes.

Nylon 6:6

Preparation

Nylon 6:6 is obtained the polymerization of adipic acid with hexamethylenediamine.

Nylon-6

It is prepared by the self-polymerization of caprolactam.

$$n \text{ caprolactam} \xrightarrow{\text{Self polymerisation}} \left[\begin{array}{c} O \\ \| \\ C \end{array} - \left[CH_2 \right]_5 - \overset{H}{N} \right]_n$$

caprolactam

Nylon-11

- It is prepared by self-condensation of ω-amino undecanoic acid.

$$n\,H_2N - (CH_2)_{10} - COOH \longrightarrow \left[\overset{H}{N} - (CH_2)_{10} - \overset{O}{\underset{\|}{C}} \right]_n$$

Properties

(i)   Nylons are translucent, whitest horny and high melting polymers.

(ii) They possess high-temperature stability and good abrasion resistance.

(iii) They are insoluble in common organic solvents and soluble in phenol and formic acid.

Uses

(i)   Nylon -6 and nylon -11 are used for moulding purposes.

(ii) Nylon 6:6 is used for fibres, which is used in making socks ,dresses, carpets etc.

Fabrication or Moulding Methods

- Shaping an article from a plastic material by the application of heat and pressure in a closed chamber is called moulding.
- The method by which plastics can be fabricated depends on the thermal behaviour of the plastics material. i.e whether it is a thermoplastic or thermosetting plastic and also on the nature of the finished product.
- The following methods are used for the fabrication of plastic material.

Moulding Methods

1.   **Hot moulding.**
   - Compressionmoulding.
   - Injection moulding.

- Transfermoulding.

2.   Extrusion moulding.

3.   Blowing.

1. *Hot Moulding*

a) *Compression Moulding*

The compression moulding is widely used to produce articles from athermoplastic material.

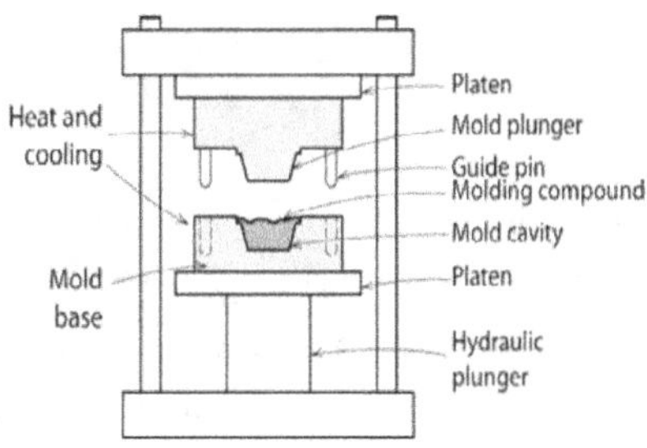

The mould is made up of two halves, Upper halves and Lower halves. Lower halves usually contain a cavity and the upper halves have a projection which fits into the cavity when the mould is closed. The gap between the projected upper half and the cavity in the lower one gives the shape of the moulded article. The thermosetting material to be moulded is placed in the cavity of themould.The plastic material is then subjected to heat and pressure is 200^0 C and 70 kg/cm^2 simultaneously.As the mould closes down under pressure , the material is compressed between the two halves and compacted to shape inside the cavity. The excess material which flows out of the mould as a thin film is known as flash. Under the influence of heat, the compacted mass gets cured and hardened to shape. The mould can be opened while it is still hot to release the moulded product.

The compression moulding is not adopted for thermoplastics because the moulded piece can be removed only after chilling the mould as the hardening of the plastic material takes place on cooling. This results in a comparatively long time for the operation with alternating and cooling.

b) Injection

A definite quantity of molten thermoplastic material is injected under pressure into a mould which is kept at thelower temperature for cooling.the molten material solidifies and takes the shape of the mould.

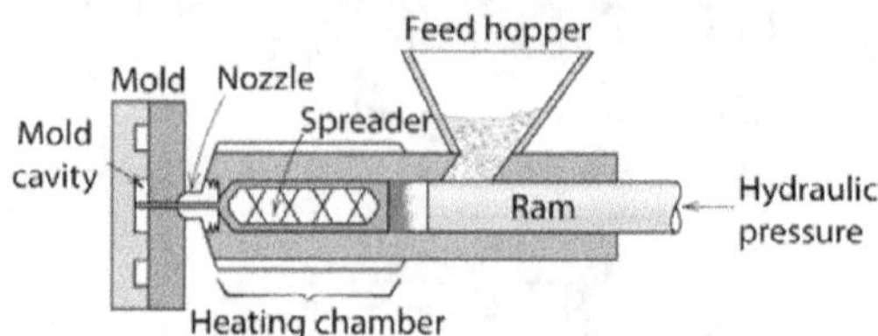

The compounded plastic material in the form of pellets or powder is fed through thehopper. The thermoplastic material softens under the influence of heat and becomes fluid. The hot fluid is forced through a nozzle by the plunger into a tightly locked mould. The mould is made up of two parts which can be separated after cooling the moulded article is released from the mould.

It is not suited for thermosetting plastic because once heated they begin to set in the reservoir.

Advantages

(1)   Low moulding and finishing cost.

(2)   High-speed production.

(3)   Low loss of materials.

2. Extrusion Moulding

This method is used for producing continuous material like sheets, tubes, rods etc.This method is only used for thermoplastics. In this process, the plastic material is fed through the hopper into the cylindrical body with electrical heating. The molten plastic materials are then pushed by means of a revolving screw arrangement into a die having the required shape of the article to be manufactured.

The finished product that extrudes out is cooled by atmospheric air. A long conveyor carries away the cooled product.

Processing Plastics – Extrusion

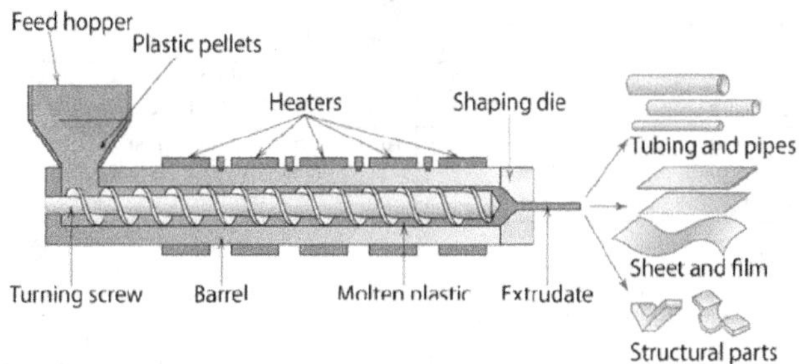

The insulated electric cables are produced by vertical extrusion moulding.

3. Blowing

Most of the hollow plastic materials like soft drink bottles, containers etc. are produced by this technique. Here a tube is made by the extrusion moulding process and theair is blown into the heated tube kept in a split mould with easily detachable halves.

The softened material gets blown into the shape of the mould.

Blow moulding

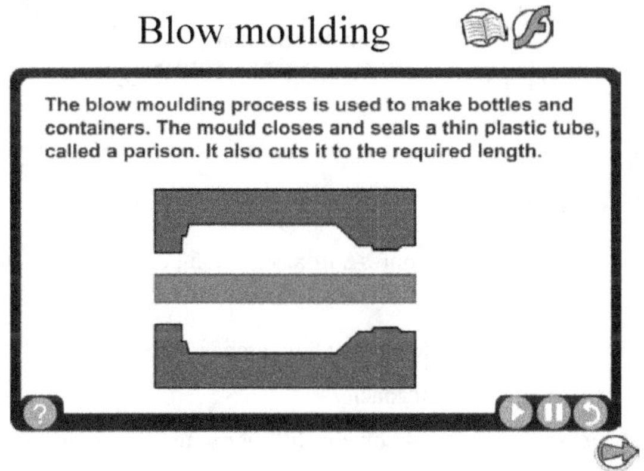

Rubber or Elastomers

Rubbers or elastomers are non-crystalline high polymers (linear polymers) having elastic and other rubber-like properties.

Types of rubbers:

Rubber can be obtained from two sources , which may be natural or artificial thus, we have,

1.	Natural rubbers.

2.	Synthetic rubber. (elastomers).

RUBBER	
Natural Rubber	Synthetic Rubber
It is obtained from the latex tree, which is a dispersion of isoprene	Synthetic rubber is obtained by polymerization of different monomers.
During the treatment of latex, an isoprene molecule undergoes polymerization to form polyisoprene Eg.: CH_2-C=C-CH_2	An elastomeris vulcanisedman-maderubber-like polymer, which is superior to natural rubber in certain properties. Eg.: SBR

Natural Rubber

Natural rubber is obtained from the tree as latex, which is   a dispersion of isoprene. During the treatment of latex, these isoprene molecules undergo polymerization to form along chain of polyisoprene.

Synthetic Rubber

An elastomer (synthetic rubber) is any vulcaniseman-maderubber-like polymer which is superior to natural rubber in certain properties.

Example: SBR

Drawbacks of Raw Rubber

- It is plastic in nature ie it becomes soft at high temperature and is too brittle at low temperature.

- It has poor strength like 200Kg/cm2 at low temperature

- It has large water absorption capacity

- It is non-resistant to non-polar solvents like benzene , vegetable and mineral oils.

- It is attacked by oxidising agents like HNO_3 and H_2SO_4

- It swells and disintegrates gradually in organic solvents.

Vulcanization of Rubber

How to improve the properties of rubber?

- The properties of rubber can be improved by compounding it with some chemicals like sulphur, hydrogen sulphide, benzoyl chloride etc, but most important addition is sulphur.

Need of Vulcanization

- Uncross-linked rubber products such as natural rubber obtained from latex are soft and have poor tensile strength and abrasion resistance. To obtain a cross-linked structure of rubber the process of vulcanization is made.

Process

- The process of vulcanization consists of heating the raw rubber with sulphur to about 100-140° C
- The added sulphur combines chemically at the double bond of different long-chain rubber springs .
- Vulcanization prevents the intermolecular movement of rubber springs.

Comparison of Rubber

Raw Rubber	Vulcanised Rubber
Tensile strength is low	Tensile strength is high
Water absorbing tendency is high	Water absorbing tendency is low
Oxidation resistance is low	Oxidation resistance is high
Low resistance to wear and tear	High resistance to wear and tear
It has high elongation	It has moderate elongation
The useful temperature is 10-60°C	The useful temperature is 40-100°C

Important Synthetic Rubbers-SBR

- An elastomer or synthetic rubber is any vulcaniseman-maderubber-like polymer which can be stretched to at least twice its original shape and dimensions as soon as thestretching force is released.
- SBR (styrene butadiene rubber) or GR-S or Buna-S.

SBR is the polymer of about 75% butadiene and 25% styrene.

Preparation

- SBR is obtained by copolymerizing an aqueous emulsion of the mixture containing 75% butadiene,25% styrene and an emulsifying agent.

$$n \; \text{(styrene)} \; + \; n \; H_2C{=}\underset{H}{C}{-}\underset{H}{C}{=}CH_2 \; \text{(butadiene)} \; \xrightarrow{\text{copolymerisation}} \; \left[\underset{}{C}{-}C{-}C{-}C{=}C{-}C\right]_n$$

Properties

- SBR is resistant to abrasion and possesses high load bearing capacity.
- It undergoes oxidation readily when a trace of ozone is present in the atmosphere.
- It requires less sulphur for vulcanization when compared to natural rubber.
- Tensile strength and flexibility of SBR are inferior to those of natural rubber.

Uses

- SBR is used for making light duty tyres, belts hoses and gaskets.
- It is also used in thefootwear industry.
- It is used as an adhesive and in electrical insulation.

Butyl Rubber (GR-I) Rubber

Butyl rubber is the copolymer of isobutylene and a small amount of isoprene.

Preparation

It is obtained by copolymerizing isobutylene with 1.5 to 4.5% isoprene in methyl chloride.

A Catalyst solution , made by dissolving anhydrous AlCl3 in methyl chloride is added to the reaction mixture.

$$n \; H_2C{=}\underset{CH_3}{\overset{CH_3}{C}} \; (\text{Isobutylene}) \; + \; n \; H_2C{=}\underset{H}{\overset{CH_3}{C}}{-}C{=}CH_2 \; (\text{Isoprene}) \; \longrightarrow \; \left[C{-}\underset{CH_3}{\overset{CH_3}{C}}{-}C{-}\underset{H}{\overset{CH_3}{C}}{=}C{-}C\right]_n \; (\text{Butyl rubber})$$

Properties

1. Butyl rubber is amorphous under normal conditions.
2. Unstabilizedpolyisobutylenes are degraded by light or heat to sticky low molecular weight products.
3. It has low permeability to gases.
4. It is soluble in hydrocarbon solvent.
5. It possesses good electrical insulating property and heat resistance and abrasion.

Uses

(1) Butyl rubber is used for making tubesand belts etc.
(2) It is also used for wire and cable insulation.

UNIT III

ELECTRO CHEMISTRY AND CORROSION SCIENCE

3.1. Introduction

Electrochemistry is a branch of chemistry, which deals with the chemical applications of electricity. Electrochemistry deals with the chemical reactions produced by passing an electric current through an electrolyte or the production of electric current through chemical reactions.

Conductors

A substance or material that allows electrical current to pass through it is called a conductor. The ability of a material to conduct electric current is called conductance. All metals, graphite, fused salts, aqueous solutions of acids, bases, etc.,

Non-Conductors (or) Insulators

Materials which do not conduct electric current are called non-conductors or insulators. *Plastics,* wood, *most of the non-metals, etc.,*

Types of Conductors

The conductors are broadly classified into two types as follows.

Metallic Conductors (or) Electronic Conductors

Metallic conductors are solid substances, which conduct electrical current due to the movement of electrons from one end to another end. The conduction decreases with the increase of temperature. *All metals, graphite.*

Electrolytic Conductors

Electrolytic conductors conduct electric current due to the movement of ions in solution or in a fused state. The conduction increases with the increase of temperature. *Acids, bases, electrovalent substances.*

Table: Differences between Metallic Conduction and Electrolytic Conduction

S.No.	Metallic conduction	Electrolytic conduction
1.	It involves the flow of electrons.	It involves the movement of ions present in the solution.
2.	It does not involve any transfer of matter.	It involves the transfer of the electrolyte.
3.	Conduction decreases with increase in temperature.	Conduction increases with increase in temperature.
4.	No changes in the chemical properties of the conductor.	Chemical reactions will take place at both the electrodes.

Cell Terminology

1. CURRENT

   Current is the flow of electrons through a wire or any conductor.

2. ELECTRODE

   Electrode is a material (or) a metallic rod/bar/stripwhich conducts electrons.

3. ANODE

   The anode is the electrode at which oxidation occurs.

4. CATHODE

   The cathode is the electrode at which reduction occurs.

5. ELECTROLYTE

   The electrolyte is a water-soluble substance forming ions in solution, and conduct an electric current.

6. ANODE COMPARTMENT

   The anodic compartment is the compartment of the cell in which oxidation half-reaction occurs. It contains the anode.

7. CATHODE COMPARTMENT

   Cathode compartment is the compartment of the cell in which reduction half reaction occurs. It contains the cathode.

8. CELL

   The cell is a device consisting two half cell. The two half cells are connected through one wire.

3.2.   Electrode Potential

A metal (M) consists of metal ions (M^{n+}) with valence electrons. When the metal (M) is placed in a solution of its own salt, any one of the following reactions will occur.

1. Positive metal ions may pass into the solution.

$$M ----->M^{n+} + ne^- \text{ (oxidation)}$$

2. Positive metal ions from the solution may deposit over the metal.

$$M^{n+} + ne^- -----> M \text{ (reduction)}$$

Examples – 1 :Zn electrode dipped in ZnSO$_4$ solution

When the Zn electrode is dipped in the $ZnSO_4$ solution, Zn goes into the solution as Zn^{2+}ions. Now, the Zn electrode attains a negative charge, due to the accumulation of valence electrons in the metal. The negative charges developed on the electrode attract the positive ions from solution. Due to this attraction, the positive ions remain close to the metal. (Fig. 4.2.a)

Example -2 Cu electrode dipped in CuSO₄ solution

When the Cu electrode is dipped in the $CuSO_4$ solution, Cu^{2+} ions from the solution deposit over the metal.

Now, the Cu electrode attains a positive charge, due to the accumulation of Cu^{2+} ions on the metal.

The positive charges developed on the electrode attract the negative ions from solution. Due to this attraction, the negative ions remain close to the metal.

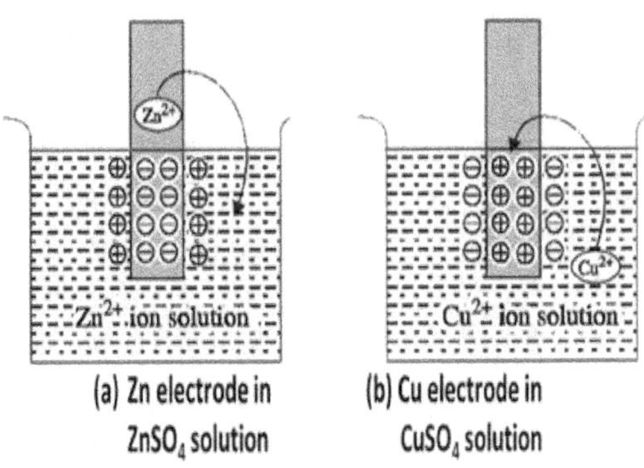

Electrode Potemtial

Thus, a sort of layer (positive (or) negative ions) is formed all around the metal. This layer is called Helmholtz electrical double layer. This layer prevents the further passing of the positive ions from or to the metal.

A difference of potential is consequently set up between the metal and the solution. At equilibrium, the potential difference becomes a constant value, which is known as the electrode potential of a metal.

Thus, the tendency of an electrode to lose electrons is called the oxidation potential, and the tendency of an electrode to gain electrons is called the reduction potential.

Factors Affecting Electrode Potential

The rate of the above reactions depends on

1. The nature of the metal.
2. The temperature.
3. The concentration of metal ions in solution.

Single Electrode Potential (E)

It is the measure of the tendency of a metallic electrode to lose or gain electrons when it is in contact with a solution of its own salt.

Standard Electrode Potential (E°)

It is the measure of the tendency of a metallic electrode to lose or gain electrons when it is in contact with a solution of its own salt of 1 molar concentration at 25°C.

Nernst Equation for Electrode Potential

Consider the following redox reaction

$$M^{n+} + n\ e^- \rightarrow M$$

For such a redox reversible reaction, the free energy change (ΔG) and its equilibrium constant (K) are interrelated as

$$\Delta G = -RT \ln K + RT \ln \frac{[\text{Product}]}{[\text{Reactant}]}$$

$$\Delta G = -\ \Delta G° + RT \ln \frac{[\text{Product}]}{[\text{Reactant}]} \quad \text{---------------- (1)}$$

$$\left[\text{ Since } -\Delta G° = -RT \ln K \right]$$

Where,

$\Delta G°$ = Standard free energy change.

The above equation (1) is known as **Van't Hoff isotherm.** The decrease in free energy ($-\Delta G$) in the above reaction involves transfer of 'n' number of electrons, then 'n' Faraday of electricity will flow. If E is the emf of the cell, then the total electrical energy (nEF) produced in the cell is

$$-\Delta G = nEF \quad (\text{or}) \quad -\Delta G° = nE°F \text{--------------------(2)}$$

Where,

$-\Delta G$ = decrease in free energy change.

(or) $-\Delta G°$ = decrease in standard free energy change. Comparing equation 1 and 2, it Becomes,

$$-nEF = -nE°F + RT \ln \frac{[M]}{[Mn^+]} \quad \text{----------------(3)}$$

Dividing the above equation (3) by – nF

$$\frac{-nEF}{nF} = \frac{-nE°F}{nF} + \frac{RT}{nF}\ln\frac{[M]}{[Mn^+]}.$$

$$-E = -E° + \frac{RT}{nF}\ln\frac{[M]}{[Mn^+]}.$$

$$E = E° - \frac{RT}{nF}\ln\frac{[M]}{[Mn^+]}.$$

[Since the activity of solid metal [M] = 1]

$$E = E° - \frac{RT}{nF}\ln\frac{1}{[Mn^+]}.$$

OR

$$\boxed{E = E° + \frac{2.303RT}{nF}\log[M^{n+}]}$$

$$E = E° + \frac{RT}{nF}\ln[Mn^+].$$

Where, R = 8.314 J/K/mole; F = 96500 coulombs;

T = 298 K (25°C), the above equation becomes,

$$\boxed{E = E°_{red} + \frac{0.0591}{n}\log[M^{n+}]}$$

In general,

$$E = E°_{oxi} + \frac{0.0591}{n}\log C$$

Similarly for oxidation potential

$$\boxed{E = E°_{oxi} + \frac{0.0591}{n}\log[M^{n+}]}$$

The above equation 5 & 6 are known as "Nernst equation for single electrode potential".

Applications of Nernst Equations

1. Nernst equation is used to calculate the electrode potential of an unknown metal.
2. The corrosion tendency of metals can be predicted.

Measurement of Single Electrode Potential

It is impossible to determine the absolute value of a single electrode potential. But, we can measure the potential difference between two electrodes potentiometrically, by combining them to form a complete cell.

For this purpose, 'reference electrode' is used. Standard hydrogen electrode (SHE) is the commonly used reference electrode, whose potential has been arbitrarily fixed as zero. The emf of the cell is measured and it is equal to the potential of the electrode. In some cases, saturated calomel electrode is used as reference electrode.

3.3. Reference Electrodes (Standard Electrodes)

The electrode potential is found out by coupling the electrode with another reference electrode, the potential of which is known or arbitrarily fixed at zero. The important primary reference electrode used is a standard hydrogen electrode, the standard electrode potential of which is taken as zero.

It is very difficult to set up a hydrogen electrode. So other electrodes called secondary reference electrodes like calomel electrodes are used.

Primary Reference Electrode (Standard Hydrogen Electrode) Construction

Hydrogen electrode consists of platinum foil, that is connected to a platinum wire and sealed in a glass tube.

Hydrogen gas is passed through the side arm of the glass tube. This electrode, when dipped in a 1N HCl and hydrogen gas at 1 atmospheric pressure is passed forms a standard hydrogen electrode. The electrode potential of SHE is zero at all temperatures. (Fig. 1.2).

It is represented as,

$$\text{Pt}, \text{H}_2(1 \text{ atm})/\text{H}^+ (1 \text{ M}); E° = 0 \text{ V}$$

In a cell, when this electrode acts as anode, the electrode reaction can be written as

$$\text{H}_2 (g) \rightarrow 2\text{H}^+ + 2e^-$$

When this electrode acts as the cathode, the electrode reaction can be written as

$$2\text{H}^+ + 2e^- \rightarrow \text{H}_2(g)$$

Limitations

1. It requires hydrogen gas and is difficult to set up and transport.

2. It requires considerable volume of test solution.

3. The solution may poison the surface of the platinum electrode.

4. The potential of the electrode is altered by changes in barometric pressure.

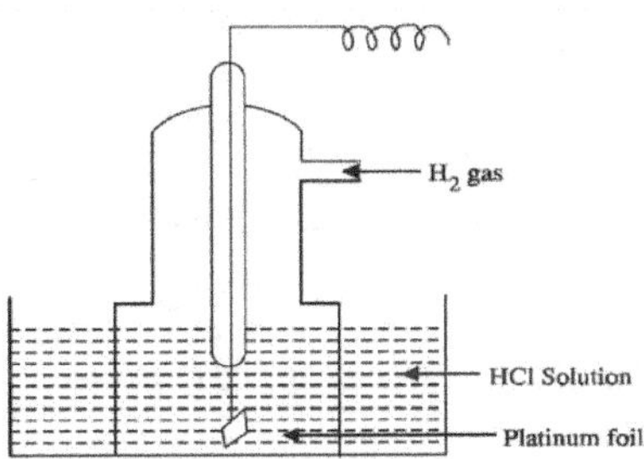

Secondary Reference Electrode

Eg.: Saturated Calomel Electrode

Construction

Calomel electrode consists of a glass tube containing mercury at the bottom over which mercurous chloride is placed.

The remaining portion of the tube is filled with a saturated solution of KCl. The bottom of the tube is sealed with a platinum wire.

The side tube is used for making electrical contact with a salt bridge. The electrode potential of the calomel electrode is + 0.2422 V.

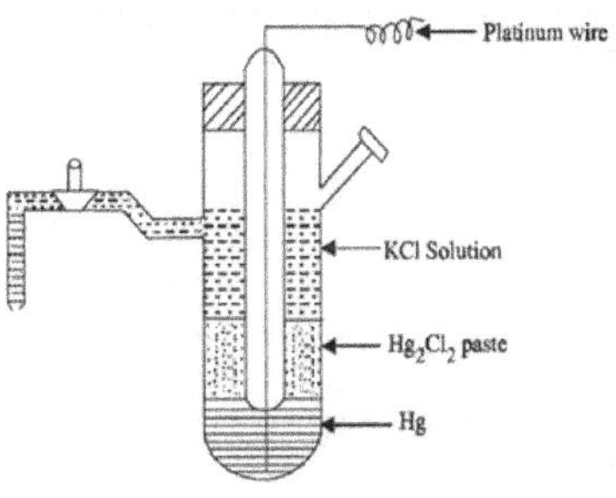

Calomel Electorde

It is represented as,

$$\text{Hg, Hg}_2\text{Cl}_2(s), \text{KCl(sat.solution); } E° = 0.2422 \text{ V}$$

If the electrode acts as anode the reaction is

$$2Hg_{(l)} \text{-----> } Hg_2^{2+} + 2e^-$$

$$Hg_2^{2+} + 2Cl^- \text{------>} Hg_2Cl_{2(s)}$$

$$2Hg_{(l)} + 2Cl^- \text{------------>} Hg_2Cl_{2(s)} + 2e^-$$

If the electrode acts as cathode the reaction is

$$Hg_2 Cl_2 (s) \text{------>} Hg_2^{2+} + 2Cl^-$$

$$Hg_2^{2+} + 2e^- \text{------>} 2Hg_{(l)}$$

$$Hg_2Cl_{2(s)} + 2e^- \text{-----> } 2Hg_{(l)} + 2Cl^-$$

The Electrode Potential is given by (for Example Cathode)

$$E_{(calomel)} = E°_{(calomel)} - \frac{RT}{2F} \ln a_{cl}^-$$

The electrode potential depends on the activity of the chloride ions and it decreases the activity of the chloride ions increases. The single electrode potential of the three calomel electrodes on the hydrogen scale at 298 K is given as

- 0.1 N KCl = + 0.3338 V
- 1.0 N KCl = + 0.2800 V
- Saturated KCl = + 0.2422 V.

Measurement of Single Electrode Potential of Zn using Saturated Calomel Electrode

The saturated calomel electrode is coupled with another Zn electrode, the potential of which is to be determined. Since the reduction potential of the coupled Zn electrode is less than E° of calomel electrode (+ 0.2422 V), the calomel electrode will act as cathode and the reaction is

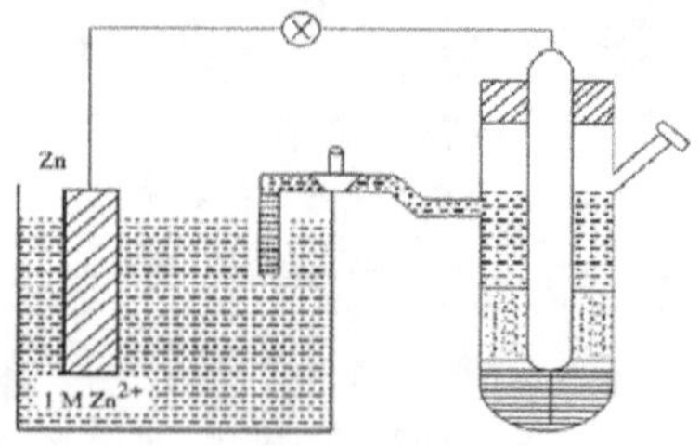

$$Hg_2Cl_{2(s)} + 2e^- \qquad 2Hg_{(l)} + 2Cl^-$$

$$E_{cell} = E°_{right} - E°_{left}$$

$$E_{cell} = E°_{cal} - E°_{Zn}$$

$$E°_{Zn} = E°_{cal} - E_{cell}$$

$$= + 0.2422 - 1.0025$$

$$E°_{Zn} = - 0.7603 \text{ volt}.$$

Ion-Selective Electrodes (ISE)

Ion-selective electrodes are the electrodes having the ability to respond only to particular ions *and* develop potential, ignoring the other ions in a mixture totally. The potential developed by an ion-selective electrode depends only on the concentration of particular ions.

Glass Electrode

The glass membrane of the glass electrode is only selective to H^+ ions only in a mixture.

Glass Electrode (Internal Reference Electrode)Construction

A glass electrode consists of thin-walled glass bulb (the glass is a special type having a low melting point and high electrical conductivity) containing a Pt wire in 0.1M HCl (Fig.1.5). The glass electrode is represented as

Pt, 0.1 M HCl / Glass

HCl in the bulb furnishes a constant H^+ ion concentration.

The glass electrode is used as the "internal reference electrode". The pH of the solutions, especially coloured solutions containing oxidising or reducing agents can be determined. The thin-walled glass bulb called glass membrane functions as an ion-exchange resin, and an equilibrium is set up between the Na^+ ions of glass and H^+ ions in solution. The potential difference varies with the H^+ ion concentration, and its emf is given by the expression

$$E_G = E°_G + 0.0592 \text{ pH}$$

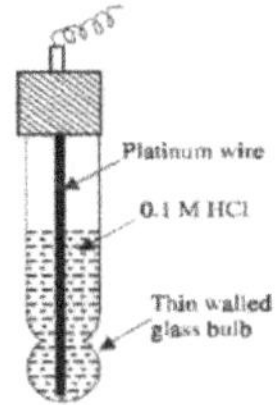

Glass Electrode

Determination of pH of a Solution using Glass Electrode

The glass electrode is placed in the solution under test and is coupled with saturated calomel electrode as shown in figure 1.6.

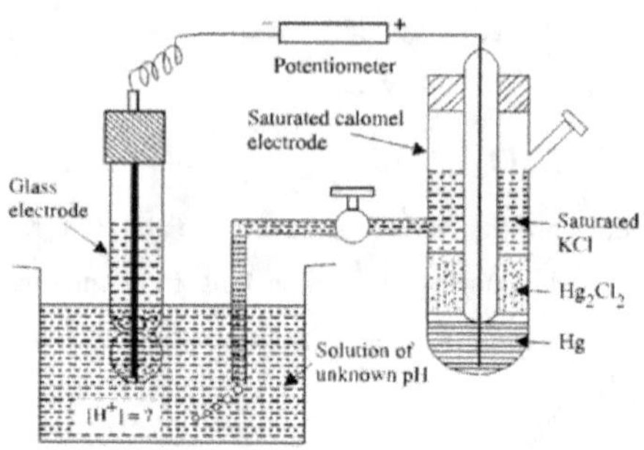

Determination of pH by Using Glass Electrode

The emf of the cell is measured.

From the emf, the pH of the solution is calculated as follows

$$E_{cell} = E_{right} - E_{left}$$

$$E_{cell} = E_{cal} - E_G$$

$$= E_{cal} - (E°_G + 0.0592\ pH)$$

$$= E_{cal} - E°_G - 0.0592\ pH$$

$$pH = \frac{E_{cal} - E°_G - E_{cell}}{0.0592} \qquad \therefore E_{cal} = 0.2422\ V$$

$$\therefore pH = \frac{0.2422 - E°_G - E_{cell}}{0.0592}$$

Advantages of Glass Electrode

1. It can be easily constructed and readily used.
2. The results are accurate.
3. It is not easily poisoned.
4. Equilibrium is rapidly achieved.

Disadvantages (Limitations)

1. Since the resistance is quite high, special electronic potentiometers are employed for measurement.

2. The glass electrode can be used in solutions only with a pH range of 0 to 10. However above the pH 12 (high alkalinity), cations of the solution affect the glass and make the electrode useless.

Applications of ISEs

1. ISEs are used in determining the concentrations of cations like H^+, Na^+, K^+, Ag^+, Li^+.
2. ISEs are used for the determination of hardness Ca^{2+} and Mg^{2+} ions).
3. Concentrations of anions like NO_3^-, CN^-, S^{2-}, halides (X^-) can be determined.
4. ISEs are used in the determination of the concentration of a gas by using gas sensing electrodes.
5. pH of the solution can be measured by using gas sensing electrode.

3.4. Types of Cells

A cell is a device consisting two half cells. Each half cell consists of an electrode dipped in an electrolytic solution. The two half cells are connected through one wire. The followings are two types of cells.

1. Electrolytic cells.
2. Electrochemical cells (or) voltaic cells (or) galvanic cells.

Electrolytic Cells

Electrolytic cells are cells in which electrical energy is used to bring about the chemical reaction. Electrolysis, electroplating, etc.

Electrochemical Cells Galvanic Cells

Electrochemical cells are entirely different from electrolytic cells. The cells used for electrolysis (where electrical energy is converted to chemical energy) are called electrolytic cells, whereas, in electrochemical cells, chemical energy is converted to electrical energy.

Galvanic cells are electrochemical cells in which the electrons, transferred due to redox reaction, are converted into electrical energy.

Cell Device (Construction)

Daniel cell consists of a zinc electrode dipped in $1MZnSO_4$ solution and a copper electrode dipped in $1M\ CuSO_4$ solution.

Each electrode is known as a half cell. The two solutions are interconnected by a salt bridge and the two electrodes are connected by a wire through the voltmeter.

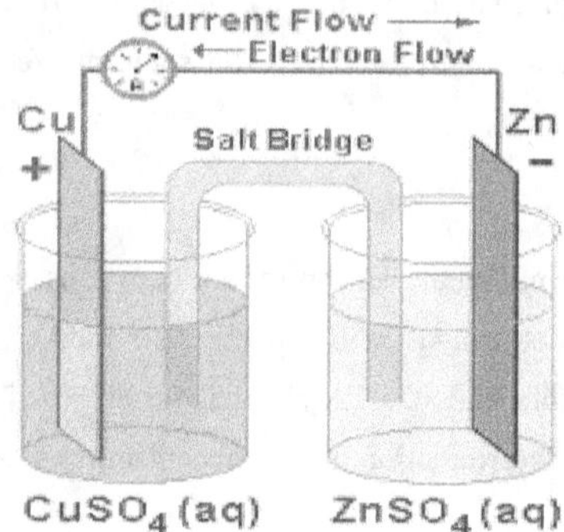

Reactions Occurring in the Cell

At anode: Oxidation takes place in the zinc electrode by the liberation of electrons, so this electrode is called negative electrode or anode.

Atcathode: Reduction takes place in the copper electrode by the acceptance of electrons, so this electrode is called the positive electrode or cathode.

$$Zn \rightarrow Zn^{2+} + 2e^- \text{ (at anode)}$$
$$Cu^{2+} + 2e^- \rightarrow Cu \text{ (at cathode)}$$
$$Cu^{2+} + Zn \rightarrow Zn^{2+} + Cu \text{ (net cell reaction)}$$

The electrons liberated by the oxidation reaction flow through the external wire and are consumed by the copper ions at the cathode.

Salt Bridge

It consists of a U-tube containing saturated solution of KCl or NaNO$_3$ in agar-agar gel. It connects the two half cells of the galvanic cells.

Functions of Salt Bridge

1. It eliminates liquid junction potential.
2. It provides the electrical continuity between the two half cells.

Conditions of a Cell to Act as Standard Cell

The conditions of an electrochemical cell to act as a standard cell are

1. The e.m.f of the cell is reproductive.
2. The temperature-coefficient of e.m.f (change in e.m.f with temperature) should be very low.

Representation of a Galvanic Cell (or) Cell Diagram

1. A galvanic cell consists of two electrodes anode and cathode.

2. The anode is written on the left-hand side while the cathode is written on the right-hand side.

3. The anode must be written by writing electrode metal first and then electrolyte. These two are separated by a vertical line or a semicolon. The electrolyte may be written by the formula of the compound (or) by ionic species.

4. The cathode must be written by writing electrolyte first and then the electrode metal. These two are separated by a vertical line or a semicolon.

5. The two half-cells are separated by a salt bridge, which is indicated by two vertical lines.

3.5. EMF of a Cell

Definition

Electromotive force is defined as, "the difference of potential which causes the flow of current from one electrode of higher potential to the other electrode of lower potential. Thus, the emf of a galvanic cell can be calculated using the following relationship.

$$EMF = \left\{\begin{array}{l}\text{Standard reduction}\\ \text{potential of right hand}\\ \text{side electrode}\end{array}\right\} - \left\{\begin{array}{l}\text{Standard reduction}\\ \text{potential of left}\\ \text{hand side electrode}\end{array}\right\}$$

$$E^{\circ}_{cell} = E^{\circ}_{right} - E^{\circ}_{left}$$

Measurement of EMF of a Cell

The potential difference or emf of a cell can be measured on the basis of Poggen Dorff's compensation principle. Here the emf of the cell is just opposed or balanced by an emf of standard cell (external emf) so that no current flows in the circuit.

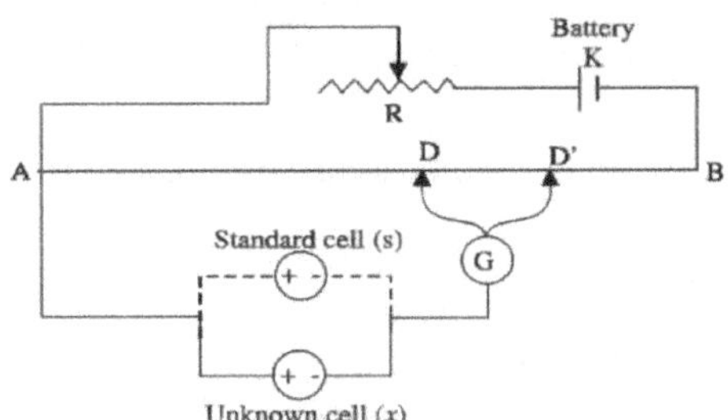

The potentiometer consists of a uniform wire AB (Fig. 1.8). A storage battery (K) is connected to the ends A and B of the wire through a rheostat (R).

The cell of unknown emf (x) is connected in the circuit by connecting its positive pole to A and the negative pole is connected to a sliding contact (D) through a galvanometer G. The sliding contact is freely moved along the wire AB till no current flows through the galvanometer. Then the distance AD is measured. The emf of the unknown cell is directly proportional to the distance AD.

Then the unknown cell (x) is replaced by a standard cell (s) in the circuit. The sliding contact is again moved till there is a null deflection in the galvanometer. Then the distance AD′ is measured. The emf of standard cell Es is directly proportional to the distance AD′.

The potentiometer consists of a uniform wire AB (Fig. 1.8). A storage battery (K) is connected to the ends A and B of the wire through a rheostat (R).

The cell of unknown emf (x) is connected in the circuit by connecting its positive pole to A and the negative pole is connected to a sliding contact (D) through a galvanometer G. The sliding contact is freely moved along the wire AB till no current flows through the galvanometer. Then the distance AD is measured. The emf of the unknown cell is directly proportional to the distance AD.

$$E_x \propto AD$$

Then the unknown cell (x) is replaced by a standard cell (s) in the circuit. The sliding contact is again moved till there is a null deflection in the galvanometer. Then the distance AD′ is measured. The emf of standard cell Es is directly proportional to the distance AD′.

$$E_s \propto AD'$$

Then, the emf of the unknown cell can be calculated from the following equation.

$$\frac{\text{Emf of the unknown cell x}}{\text{Emf of the standard cell s}} = \frac{\text{Length AD}}{\text{Length AD'}}$$

$$\frac{E_x}{E_s} = \frac{AD}{AD'} \qquad \therefore \text{Emf of the unknown cell} = E_x = \frac{AD}{AD'} \times E_s$$

Factors Affecting EMF of a Cell

1. Nature of the electrolytes and electrodes.
2. Concentration and composition of the electrolytes.
3. pH and temperature of the solution.

Applications of EMF Measurements

1. Determination of standard free energy change and equilibrium constant.
2. Determination of pH by using a standard hydrogen electrode.
3. The solubility of a sparingly soluble salt can be determined.
4. Valency of an ion can be determined.
5. Potentiometric titrations can be carried out.
6. Hydrolysis constant can also be determined.

Differences between Electrolytic Cells and Electrochemical Cells

S.No.	Electrolytic Cell	Electrochemical Cell
1.	Electrical energy is converted into electrical energy.	Chemical energy is converted into electrical energy.
2.	The anode carries a positive charge.	The anode carries a negative charge.
3.	The cathode carries a negative charge.	The cathode carries a positive charge.
4.	Here the electrons are supplied to the cell from the external battery. i.e., electrons move in through cathode and comes out from the cathode	But the electrons are drawn from the cell. i.e., electrons move from anode to cathode through the external circuit.
5.	The amount of the electricity passed during electrolysis is measured by a coulometer.	The e.m.f. produced in the cell is measured by a potentiometer.
6.	The extent of chemical reactions occurring at the electrodes is governed by Faraday's law of electrolysis.	The e.m.f. of the cell depends on the concentration of the electrolytes and the chemical nature of the electrode.
7.	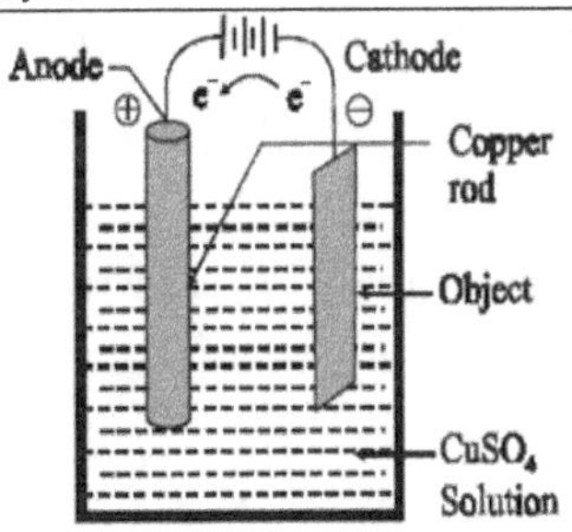 Anode — Cathode — Copper rod — Object — CuSO$_4$ Solution	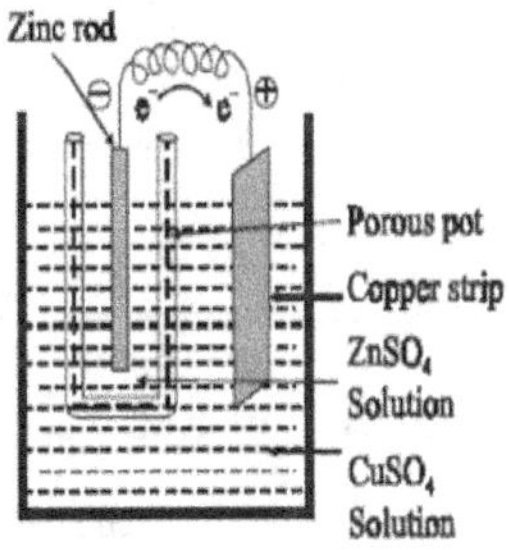 Zinc rod — Porous pot — Copper strip — ZnSO$_4$ Solution — CuSO$_4$ Solution

3.6. Reversible and Irreversible Cells

Reversible Cells

*Daniel cell, secondary batteries (rechargeable batteries).*Daniel cell is a very good example for a reversible cell. Its emf is 1.1 volt. It is represented as

Zn/ZnSO$_4$(1 M)//CuSO$_4$ (1 M)/Cu.

A cell which obeys the following three conditions of thermodynamic reversibility is called the reversible cell.

1. If the daniel cell is connected to an external source of emf equal to 1.1 volts, no current flows and also no chemical reaction takes place in the cell.

2. If the external emf is made slightly less than 1.1 volts, a few current flows from the cell and small chemical reaction occurs.

3. If the external emf is made slightly greater than 1.1 volts, the current will flow in the opposite direction. Copper will pass into the solution as copper ions and zinc will get deposited on the zinc electrode.

Irreversible Cells

Zinc – silver cell, Dry cell (Primary Cells)

Cells which do not obey the conditions of thermodynamic reversibility are called irreversible cells.

Zinc-Silver cell is an example for an irreversible cell. It is represented as

$$Zn/H_2SO_4(aq)/Ag$$

The cell reactions occur at anode and cathode are,

$$Zn + H_2SO_4 \text{ ---> } ZnSO_4 + H_2\uparrow \text{ (at anode)}$$

$$2Ag^+ + 2e^- \text{ ---> } 2Ag \text{ (at cathode)}$$

When the two electrodes are connected, zinc dissolves with the liberation of hydrogen gas.

When the external emf, slightly greater than the actual emf of the cell, is applied to it, the above reactions are not reversed.

Because one of the products, H2 gas, is already escaped. Such a cell, which does not obey the conditions of thermodynamic reversibility, is called an irreversible cell.

3.7. Corrosion and Its Control

Metals and alloys are generally used as a fabrication or construction materials engineering. If the metal or alloy structures are not properly maintained, they deteriorate slowly by the action of atmospheric gases, moisture and other chemicals. This phenomenon of deterioration or destruction of metals and alloys is known as corrosion.

Causes of Corrosion

Metals occur in nature in two different forms.

1. Native state.
2. Combined state.

1. Native State

The metals occur in native (or) free (or) uncombined state are non-reactive with the environment.

They are noble metals exist as such in the earth's crust. They have very good corrosion resistance.

Eg. Au, Pt, Ag

2. Combined State

Except noble metals, all other metals are reactive and react with the environment and form stable compounds, as their oxides, sulphides, chlorides and carbonates. They exist in their form of stable compounds called ores and minerals.

Eg. Fe_2O_3, ZnO, PbS, $CaCO_3$ etc.,

How and Why Corrosion Occurs

The metals are extracted from these compounds (ores).

During the extraction, these ores are reduced to their metallic states. In the pure metallic state, the metals are unstable as they are considered in excited state ie., higher energy state.

Therefore, as soon as the metals are extracted from their ores, the reverse process begins and formmetal compounds, which are thermodynamically stable, ie., lower energy state.

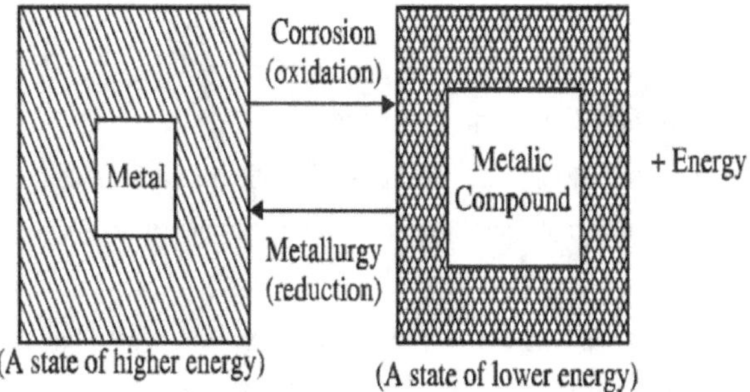

Hence, when metals are used in various forms,they are exposed to the environment, (such as dry gases, moisture, etc.,) the exposed metal surface begins to decay ie., conversion into the more stable compound. This is the basic reason for metallic corrosion.

Due to corrosion, some useful properties of metals such as electrical conductivity, ductility and malleability etc., are lost.

Consequences of Corrosion

- Due to the formation of corrosion product over the machinery, the efficiency of the machine gets lost.
- The products get contaminated due to corrosion.
- The corroded equipment must be replaced frequently.
- The plant gets a failure due to corrosion.
- It is necessary for over design to compensate for the corrosion.
- Corrosion releases toxic products, health hazard, etc.

Classification Theories of Corrosion

Based on the environment, corrosion is classified into:

- Dry or Chemical corrosion.
- Wet or Electrochemical corrosion.

Dry Chemical Corrosion

Dry corrosion is due to the attack of metal surfaces by the atmospheric gases such as oxygen, hydrogen sulphide,sulphur dioxide, nitrogen, etc.

There are 3 main types of dry corrosion.

1. Oxidation corrosion (or) corrosion by oxygen.
2. Corrosion by hydrogen.
3. Liquid-metal corrosion.

Oxidation Corrosion (or) Corrosion by Oxygen

Corrosion by oxygen oxidation, corrosion is brought about by the direct attack of oxygen at low or high temperatures on metal surfaces in the absence of moisture. Alkali metals (Li, Na, K, etc.)

Alkaline-earth metals (Mg, Ca, Sn, etc.) are rapidly oxidised at low temperature. At high temperature, almost all metals (except, Ag, Au and Pt) are oxidised.

Mechanism of Dry Corrosion

Oxidation occurs first at the surface   of the metal resulting in the formation of metal ions (M^{2+}), which occurs at the metal / oxide interface.

$$M \text{-----} > M^{2+} + 2e^-$$

Oxygen changes to ionic form (O^{2-}) due to the transfer of electron from metal, which occurs at the oxides film / environment interface

$$\tfrac{1}{2}\,O_2 + 2e^- \; -----> \; O^{2-}$$

Oxide ions react with the metal -into form the metal oxide film.

$$M + \tfrac{1}{2}\,O_2 \; -----> \; M^{2+} + O^{2-} \equiv MO \;(\text{Metal-oxide film})$$

Once the metal surface is converted to a monolayer of metal oxide, for further corrosion (oxidation) to occur, the metal ion diffuses outward through the metal-oxide barrier.

Thus, the growth of oxide film commences perpendicular to the metal surface.

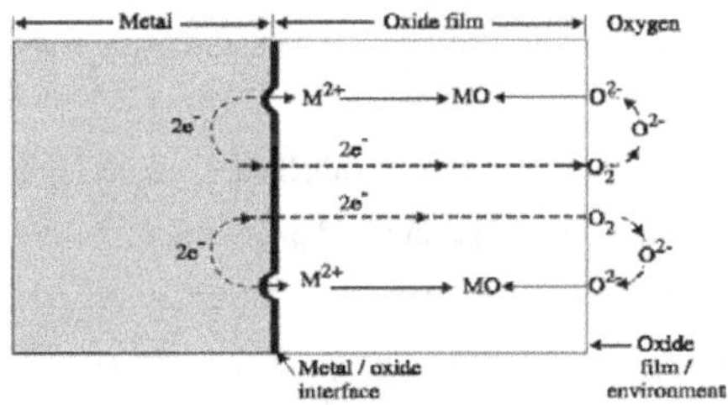

Nature of Oxide Film

The nature of oxide film formed on the metal surface plays an important role in oxidation corrosion.

Stable Oxide Layer

A stable oxide layer is a fine-grained in structure and gets adsorbed tightly to the metal surface.

Such a layer is impervious in nature and stops further oxygen attack through diffusion. Such a film behaves as a protective coating and no further corrosion can develop.

Oxides of Al, Sn, Pb, Cu, etc., are stable oxide layers.

Unstable Oxide Layer

The unstable oxide layer is mainly produced on the surface of noble metals, which decomposes back into the metal and oxygen.

$$\text{Metal Oxide} \rightarrow \text{Metal} + \text{Oxygen}$$

Eg. Oxides of Pt, Ag, etc., are unstable oxide layers.

Volatile Oxide Layer

The oxide layer volatilizes as soon as it is formed, leaving the metal surface for further corrosion.

Molybdenum oxide (MoO_3) is volatile.

Protective (or) Non-Protective Oxide Film (Pilling- Bedworth Rule)

(a) According to Pilling-Bedworth rule, if the volume of the oxide layer formed is less than the volume of the metal, the oxide layer is porous and non-protective.

(b) The volume of oxides of alkali and alkaline earth metals such as Na, Mg, Ca, etc., is less than the volume of the metal consumed. Hence, the oxide layer formed is porous and non-protective.

(c) On the other hand, if the volume of the oxide layer formed is greater than the volume of the metal, the oxide layer is non-porous and protective.

Eg. The volume of oxides of heavy metals such as Pb, Sn, etc., is greater than the volume of the metal. Hence, the oxide layer formed is non-porous and protective.

3.8. Protective Coatings

Protectivecoatings used to protect the metals from corrosion. The protective coating acts as a physical barrier between the coated metal surface and the environment. However, they are also used as a decorative purpose.

In addition to the corrosion protection and decoration, they impart some special properties such as hardness, electrical properties, oxidation resistance and thermal insulating properties to the protected surface.

Types of Protective Coating

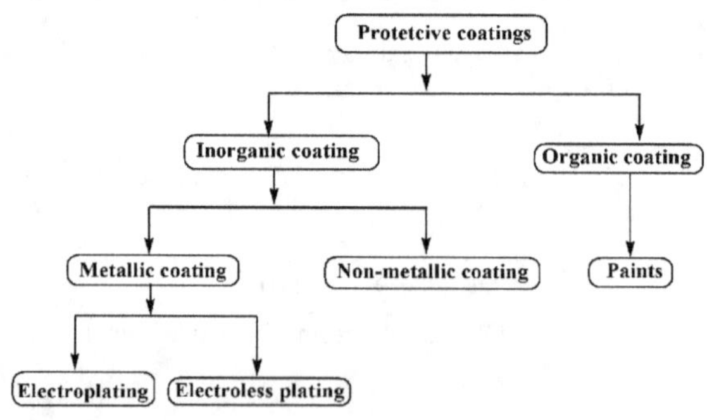

Metallic Coatings

Electroplating or Electro-Deposition

Principle

The basic principle of electroplating is coating the coating material on the base metal by passing a direct current through an electrolytic solution containing the soluble salt of the coating material.

The base metal to be plated is made cathode of an electrolytic cell, whereas the anode is either made of the coating metal itself or an inert material with good electrical conductivity.

Definition

The basic principle of electroplating is coating the coating material on the base metal by passing a direct current through an electrolytic solution containing the soluble salt of the coating material.

Objective of electroplating.

On Metals

1. To increase the resistance to corrosion of the coated metal.
2. To improve the hardness and physical appearance of the article.
3. To increase the decorative and commercial value of the article.
4. To increase the resistance of chemical attack.
5. To improve the properties of the surface of the article.

On Non-Metals

1. To increase strength.
2. To preserve and decorate the surfaces of non-metals like plastics, wood, glass etc.,
3. For making the surface conductivity by utilisation of lightweight, non-metallic materials.

Theory

During electroplating, the concentration of the electrolyte solution should remain unaltered.

This is possible only in any one of the following ways.

1. If the anode is made of coating metal itself in the electrolytic cell, during electrolysis, the concentration of electrolytic bath remains unaltered, since the metal ions deposited

from the bath on cathode are replenished continuously by the reactions of free anions with the anode.

2. If the anode is made of an inert material like graphite, electrolyte (salt of coating metal) should be added continuously to maintain the concentration of coating metal ions in the bath.

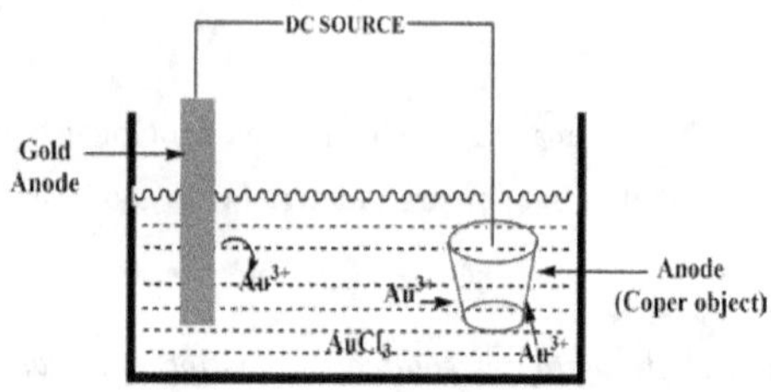

Electroplating

Process

- The copper object, to be plated, is first treated with dil. HCl or dil. H_2SO_4.
- The cleaned object is then made cathode of an electrolytic cell and gold foil as the anode.
- The $AuCl_3$ solution is taken as the electrolyte.
- When the current is passed through the solution from the battery through the solution, gold dissolves in the electrolyte and deposits uniformly on the copper object.

Chemical Reactions Involved in the Process

Electrolyte $AuCl_3$ ionises as

$$AuCl_3 \rightarrow Au^{3+} + 3Cl^-$$

At Cathode: On passing current, Au^{3+} ions moves to the cathode and get deposited there as Au metal.

$$Au^{3+} + 3e^- \rightarrow Au$$

At Anode: The free chloride ions migrate to the gold anode and dissolves an equivalent amount of Au from $AuCl_3$. In order to get strong, adherent and smooth deposit certain additive (glue, gelatin, etc) are added to the electrolytic bath. To improve the brightness of deposit, brightening agents are added in the electrolytic bath. The favourable conditions for a good electrodeposits are optimum temperature ($60°C$), optimum current density ($1\text{-}10$ mA/cm^2) and low metal ion concentrations.

Characteristics of Gold Plating

- The deposits of gold are used for electrical and electronic applications.
- It is used for high-quality decorations and high oxidation resistant coatings.
- For jewellery, thevery thin coating is given (0.05 – 1.0 microns).

Electroless Plating

Principle

Electroless plating is a technique of depositing a noble metal (from its salt solution) on a catalytically active surface of the metal, to be protected , by using a suitable reducing agent without using electrical energy.

The reducing agents reduced the metallic ions to metal which gets plated over the catalytically activated surface giving a uniform thin coating.

Metal ions + Reducing agents → Metal (deposited) + oxidised products

Electroless Nickel Plating

Step I: Pretreatment and activation of the surface

The surface to be plated is first decreased by using organic solvents or alkali, followed by acid treatment. Activation depends on the type of metal or alloy or non-metal used.

Examples:

1. The surface of stainless steel is activated by dipping in a hot solution of 50% dil. H_2SO_4.
2. The surface of Mg alloy is activated by a thin coating of Zn and Cu over it.
3. Metals and alloys like Al, Cu, Fe, Brass, etc., can be directly nickel plated without activation.
4. Nonmetallic materials are activated dipping them in a solution of containing $SnCl_2$+ HCl, followed by dipping in $PdCl_2$ solution. On drying a thin layer of Pd is formed over the surface.

Step II: Preparation of plating bath

The plating bath consists of the following ingredients.

Nature of the compound	Name of the compound	Quantity (g/l)	Function
Coating solution	$NiCl_2$	20	Coating metal
Reducing agent	Sodium hypophosphite	20	Metal ions reduced
Complexing agent cum exultant	Sodium succinate	15	Improves the quality
Buffer	Sodium acetate	10	Control the pH
Optimum pH	4.5	-	-
Optimum temperature	93ºC	-	-

Step III : Procedure for plating

The pretreated object is immersed in the plating bath for the required time. During which of the following reduction reaction will occur and Ni gets coated over the object.

Chemical reaction takes place.

$$At\ Cathode:\ Ni^{2+}\ +\ 2e^- \rightarrow\ Ni$$

$$At\ Anode:\ \quad H_2PO_2^-\ +\ H_2O\ \rightarrow\ H_2PO_3^-\ +2H^+\ +\ 2e^-$$

$$Overall\ Reaction:\ Ni2+\ +\ H_2PO_2^-\ +H_2O\ \rightarrow\ Ni\ +\ H_2PO_3^-\ +2H^+$$

Applications

1. Extensively used in electronic applications.
2. Used in domestic as well as the automotive industry.
3. Polymers can be coated and used in decorative and functional works.
4. Electroless Ni & Cu coated plastic cabinets are used in digital as well as electronic instruments.

Advantages

1. No electricity is required.
2. Electroless plating on insulators and semiconductors can be easily carried out.
3. Complicated parts can also be plated uniformly.
4. Electroless coating possesses good mechanical, chemical and magnetic properties.

Difference between Electroplating and Electroless Plating

S.No.	Electroplating	Electroless plating
1.	Carried out by passing current	Carried out by autocatalytic method
2.	Separate anode is employed	The catalytic surface of the substrate acts as anode
3.	Anodic reaction: $M \rightarrow M^{n+} + ne^-$	Anodic reaction: $R \rightarrow O + ne^-$
4.	The object to be coated is cathode	The object to be coated, after making its surface catalytically active
5.	Cathodic reaction is $M^{n+} + ne^- \rightarrow M$	Cathodic reaction is $M^{n+} + ne^- \rightarrow M$
6.	It is not satisfactory for the object having irregular shape	It is satisfactory for all parts
7.	Carried out on conducting materials	Carried out on conducting and semiconducting materials.
8.	Thickness of plating is $1 - 100\ \mu m$	Thickness of plating is $1 - 100\ \mu m$

NON-CONVENTIONAL ENERGY SOURCES

4.1.   Nuclear Fission

When U is bombarded by a thermal neutron (low energy,neutron), it splits into two approximately equal parts with the liberation of a large amount of energy.

Definition

Nuclear fission is defined as "the process of splitting of a heavier nucleus into two (or) smaller nuclei with the simultaneous liberation of large amounts of energy".

Mechanism of Nuclear Fission

When U^{235} is bombarded by thermal neutron (slow moving), unstable U^{236} is formed. The unstable U then divide into two approximately equal nuclei with the release of neutrons and a large amount of energy.

$$U^{235}_{92} + n^{1}_{0} \longrightarrow \left[U^{236}_{92}\right] \begin{cases} \longrightarrow Ba^{140}_{56} + Kr^{93}_{36} + 3n^{1}_{0} \\ \\ \longrightarrow Xe^{144}_{54} + Sr^{90}_{38} + 2n^{1}_{0} \\ \\ \longrightarrow Cs^{144}_{55} + Rb^{90}_{37} + 2n^{1}_{0} \end{cases}$$

(Unstable)

Illustration

Splitting of U235 has been shown below.

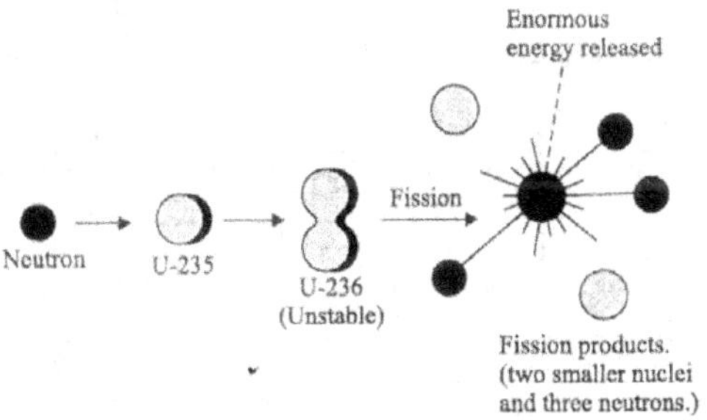

The Fission Process Illustrated

During the nuclear fission, a large amount of energy is released.

4.2. Characteristics of Nuclear Fission

- A heavy nucleus (U^{235} (or) Pu^{239}), bombarded by slow moving neutrons, split into two or more nuclei.

- Two or more neutrons are produced by fission of each nucleus.

- Large quantities of energy are produced as a result of the conversion of a small mass of the nucleus into energy.

- All the fission fragments are radioactive, giving off β and γ-radiations.

- The atomic weights of fission product range from about 70 to 160.

- All the fission reactions are a self-propagating chain-reactions because fission products contain neutrons (secondary neutrons) which further cause fission in other nuclei

- The nuclear chain reactions can be controlled and maintained steadily by absorbing a desired number of neutrons. This process is used in a nuclear reactor.

- Every secondary neutron, released in the fission process, does not strike a nucleus, some escape into the air and hence, a chain reaction cannot be maintained.

- Multiplication factor: The number of neutrons, resulting from a single fission, is known as the multiplication factor. When the multiplication factor is less than 1, a chain reaction does not take place.

Nuclear Fusion

Nuclear fusion is defined as, "the process of combination of lighter nuclei into heavier nuclei, with the simultaneous liberation of large amounts of energy". Nuclear fusion occurs in the sun.

$$_1H^2 + {}_1H^2 \longrightarrow {}_2He^4 + \text{energy}$$

S. No	Nuclear fission	Nuclear fusion
1.	It is the process of breaking a heavier nucleus.	It is the process of combining a lighter nucleus.
2.	It emits radioactive rays.	It does not emit any radioactive rays.
3.	It occurs at ordinary temperature.	It occurs at high temperature ($>10^6$K).
4.	The mass number and an atomic number of new elements are lower than that of parent nucleus.	The mass number and an atomic number of the productare higher than that of starting elements.
5.	It gives rise to the chain reaction.	It does not give rise to the chain reaction.
6.	It emits neutrons.	It emits positrons.
7.	It can be controlled.	It cannot be controlled.

Nuclear Chain Reactions

In the nuclear fission reaction, the neutrons emitted from the fission of U^{235} atom may hit other U^{235} nuclei and cause fission producing more neutrons and so on. Thus, a chain of self-sustaining nuclear reactions will be set up with the release of an enormous amount of energy. But the amount of energy released will be less than expected. Thus, the fission of U by slow moving neutrons is a chain reaction.

Definition

A fission reaction, where the neutrons from the previous step continue to propagate and repeat the reaction is called a nuclear chain reaction.

Reason for Less Energy

Some of the neutrons, released in the fission of U^{235} may escape from the surface to the surroundings or may be absorbed by U^{238} present as an impurity. This will result in the breaking of the chain and the amount of energy released will be less than expected.

Criteria for Nuclear Chain Reaction

For a nuclear chain reaction to continue, sufficient amount of U^{235} must be present to capture the neutrons, otherwise, neutrons will escape from the surface.

Critical Mass

The minimum amount of fissionable material (U^{235}) required continuing the nuclear chain reaction is called critical mass

a) Supercritical Mass

If the mass of the fissionable material (U^{235}) is more the critical mass, it is called supercritical mass.

b) Sub-Critical Mass

If the mass of the fissionable material is smaller than the critical mass, it is called sub-critical mass.

Thus, the mass greater or lesser than the critical mass will hinder the propagation of the chain reaction.

Illustration

When U^{235} nucleus is hit by a thermal neutron, it undergoes the following reaction with the release of three neutrons.

$$U_{92}^{235} + n_0^1 \longrightarrow Ba_{56}^{139} + Kr_{36}^{94} + 3n_0^1$$

Each of the three neutrons, produced in the above reaction, strikes another U^{235} nucleus causing (3x3) 9 subsequent reactions. These 9 reactions further give rise to (3 x 9) 27 reactions.This process of propagation of the reaction by multiplication in threes at each fission is called a chain reaction.

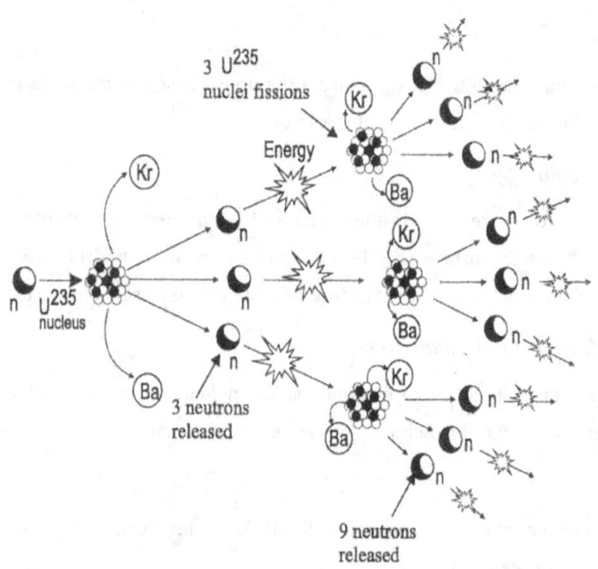

U^{235} Fission Chain Reaction Illustrated

Nuclear Energy

The enormous amount of energy released during the nuclear chain reaction of the heavy isotope like U^{235} (or) Pu^{239} is called nuclear energy.

Definition

The energy released by the nuclear fission is called nuclear fission energy (or) nuclear energy.

Illustration

The fission of U or Pu^{239} occurs instantaneously, producing an enormous amount of energy in the form of heat and radiation.

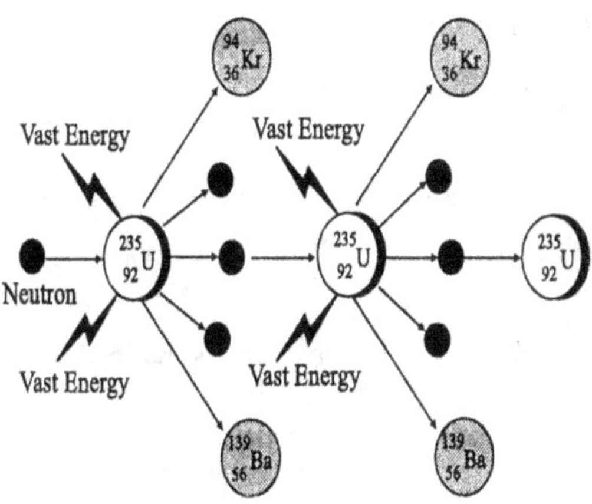

Nuclear Energy Illustrated

Cause of the Release of Energy

The enormous amount of energy released during the nuclear fission is due to the loss of some mass when the reaction takes place. It has been observed that during nuclear fission, the sum of the masses of the products formed is slightly less than the sum of masses of target species and bombarding neutron. The loss in mass gets converted into energy according to Einstein equation

$$E = mc^2$$

Where

c = velocity; m = loss in mass and E = energy.

Nuclear Reactor or Pile

If a nuclear fission reaction is made to occur in a controlled manner, then the energy released can be used for many constructive purposes.

Definition

The arrangement or equipment used to carry out the fission reaction under controlled conditions is called a nuclear reactor.

Example

The energy released (due to the controlled fission of U^{235} in a nuclear reactor) can be used to produce steam, which can run turbines and produce electricity.

Components of a Nuclear Reactor

The main components of the nuclear reactor are

1. Fuel Rods

The fissionable materials used in the nuclear reactor are enriched U^{235}. The enriched fuel is used in the reactor in the form of rods or strips.

Example : U^{235}, P^{239}

Function: It produces heat energy and neutrons, that starts a nuclear chain reaction.

2. Control Rods

To control the fission reaction (rate), movable rods, made of cadmium (or) boron, are suspended between fuel rods. These rods can be lowered or raised to control the fission reaction by absorbing excess neutrons. If the rods are deeply inserted into the reactor, they will absorb more neutrons and the reaction becomes very slow. On the other hand, if the rods are pushed outwards, they will absorb fewer neutrons and the reaction will be very fast.

$$Cd_{43}^{113} + n_0^1 \longrightarrow Cd_{43}^{114} + \gamma\text{-ray}$$

$$B_5^{10} + n_0^1 \longrightarrow + B_5^{11} + \gamma\text{-ray}$$

Example : Cd^{113}, B^{10}

Function: It controls the nuclear chain-reaction and avoids the damage to the reactors.

3. Moderators

The substances used to slow down the neutrons are called moderators.

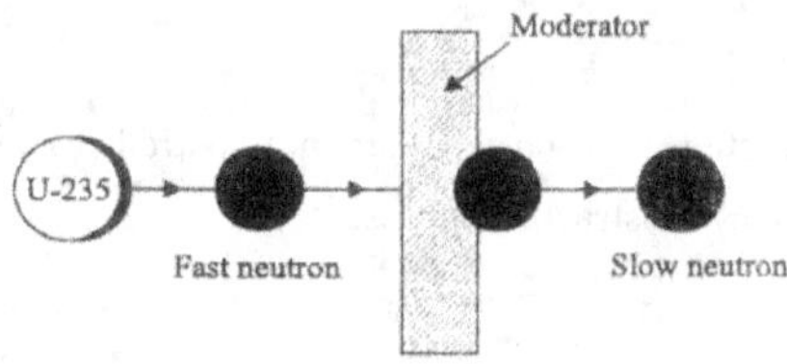

Functions of a Moderator

When the fast-moving neutrons collide with the moderator. They lose energy and gets slow down.

Example: *Ordinary water, heavy water, graphite, beryllium*

Function: The kinetic energy of fast moving neutrons(1 meV) is reduced to slow neutrons (0.25 eV).

4. Coolants

In order to absorb the heat produced during fission, a liquid called coolant is circulated in the reactor core. It enters the base of the reactor and leaves at the top. The heat carried by the out-going liquid is used to produce steam.

Example: Water **(act** as moderator & coolant), heavy water, liquid metal (Na or K), air (CO_2).

Function: It cools the fuel core.

5. Pressure Vessel

It encloses the core and also provides the entrance and exit passages for coolant.

Function: It withstands the pressure as high as 200 kg/cm².

6. Protective Shield

The nuclear reactor is enclosed in a thick massive concrete shield (more **than 10** meters thick).

Function: The environment and operating personnel are protected from destruction in case of leakage of radiation.

4.3. Light Water Nuclear-Power-Plant

Definition

Light-water nuclear-power plant is the one, in which U^{235} fuel rods are submerged in water. Here the water acts as coolant and moderator

Working

The fission reaction is controlled by inserting or removing the control rods of B^{10} automatically from the spaces in between the fuel rods. The heat emitted by fission of U^{235} in the fuel core is absorbed by the coolant (light water). The heated coolant (water at 300°C) then goes to the heat exchanger containing sea water. The coolant here transfers heat to sea water, which is converted into steam. The steam then drives the turbines, generating electricity.

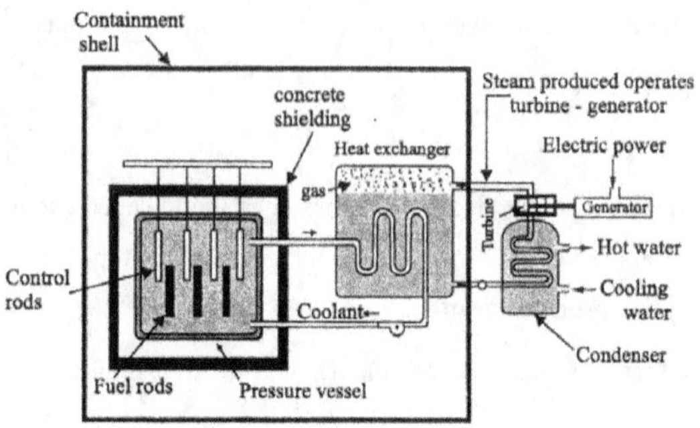

Light Water Nuclear Power Plant

Pollution

Though nuclear power plants are very important for the production of electricity, they will cause a serious danger to environments.

Problem of Disposal of Reactor Waste

Disposal of reactor waste is another important problem because the fission products viz., Ba^{139} & Kr^{92} is themselves radioactive. They emit dangerous radiation for several hundred years. So the waste is packed in concrete barrels, which are buried deep in the sea.

4.4. Breeder Reactor

Breeder reactor is the one which converts non-fissionable material (U^{238}, Th^{232}) into the fissionable material (U^{235}, Pu^{239}). Thus, the reactor produces or breeds more fissionable material than it consumes.

$$\underset{\text{Non–fissionable}}{U_{92}^{238}} + n_0^1 \longrightarrow + \underset{\text{Fissionable}}{Pu_{94}^{239}} + 2e^-$$

$$Pu_{94}^{239} + n_0^1 \longrightarrow \text{Fission products} + 3n_0^1$$

In a breeder reactor, of the three neutrons emitted In the fission of U^{235}, only one is used in propagating the fission chain with U^{235}. The other two are allowed to react with U^{238}. Thus, two fissionable atoms of Pu^{239} are produced for each atom of U^{235} consumed. Therefore, the breeder reactor produces more fissionable material than it uses. Hence Pu^{239} is a man-made nuclear fuel and is known as secondary nuclear fuel.

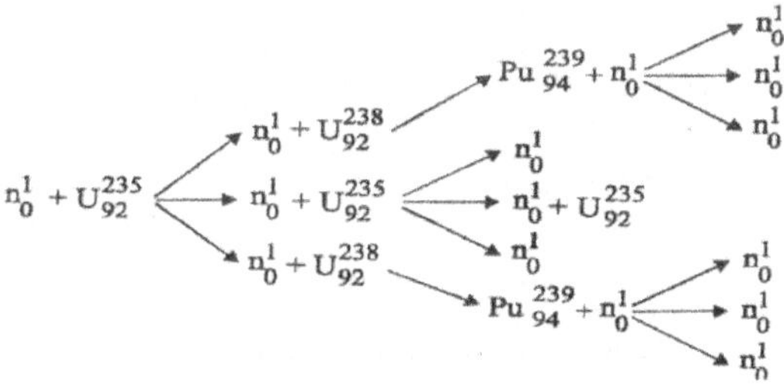

Principle of Breeder Reactor

Significance

- The non-fissionable nuclides, such as U^{238} & Th^{232}, called fertile nuclides are converted into fissile nuclides

- The fissionable nuclides such as U^{235} & Pu^{239} are called fissile nuclides.

- As the regeneration of fissile nuclei takes place, its efficiency is more.

4.5. Solar Energy Conversion

Solar energy conversion is the process of conversion of direct sunlight into more useful forms. This solar energy conversion occurs by the following two mechanisms.

1. Thermal conversion.
2. Photoconversion.

Thermal Conversion

Thermal conversion involves absorption of thermal energy in the form of IR radiation. Solar energy is an important source for low-temperature heat (temperature below 100°C), which is useful for heating buildings, water and refrigeration. Methods of thermal conversion

1. Solar heat collectors.
2. Solar water heater.

1. Solar Heat Collectors

Solar heat collectors consist of natural materials like stones, bricks (or) materials like glass, which can absorb heat during the daytime and release it slowly at night.

Uses

It is generally used in cold places, where houses are kept in hot condition using solar heat collectors.

2. Solar Water Heater

It consists of an insulated box inside of which is painted with black paint. It is also provided with, a glass lid to receive and store solar heat. Inside the box, it has a black painted copper coil, through which cold water is allowed to flow in, which gets heated up and flows out into a storage tank. From the storage tank water is then supplied through pipes.

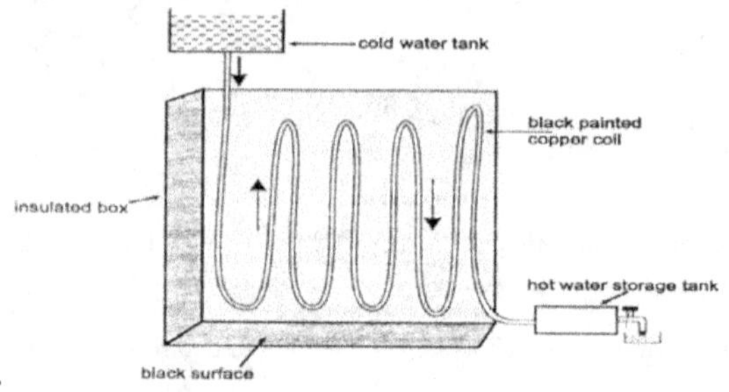

Solar Water Heater

Photoconversion

Photoconversion involves the conversion of light energy directly into electrical energy.

Methods of Photoconversion

Photo conversion can be made by the following method.Photogalvanic cell (or) Solar cell.

4.6. Photogalvanic Cell or Solar Cell

Definition

The photogalvanic cell is the one, which converts the solar energy (energy obtained from the sun) directly into electrical energy.

Principle

The basic principle involved in the solar cells is based on the photovoltaic (PV) effect. When the solar rays fall on a two layer of semiconductor devices, a potential difference between the two layers is produced. This potential difference causes the flow of electrons and produces electricity.

Construction

Solar cells consist of a p-type semiconductor (such as Si doped with B) and an n-type semiconductor (such as Si doped with P). They are in close contact with each other.

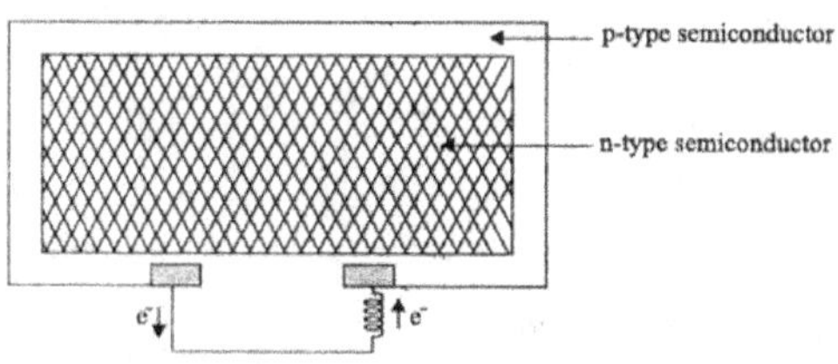

Solar Cell

Working

When the solar rays fall on the top layer of p-type semiconductor, the electrons from the valence band get promoted to the conduction band and cross the p-n junction into the n-type semiconductor.

Therebypotential difference between the two layers is created, which causes the flow of electrons (ie., an electric current). The potential difference and hence current increases as more solar rays fall on the surface of the top layer.

Thus, when this p and n layers are connected to an external circuit, electrons flow from n-layer to p-layer, and hence current is generated.

Applications of Solar Cells

1. Lighting Purpose

Solar cells can be used for lighting purpose. Nowadays,electrical street lights are replaced by solar street lights.

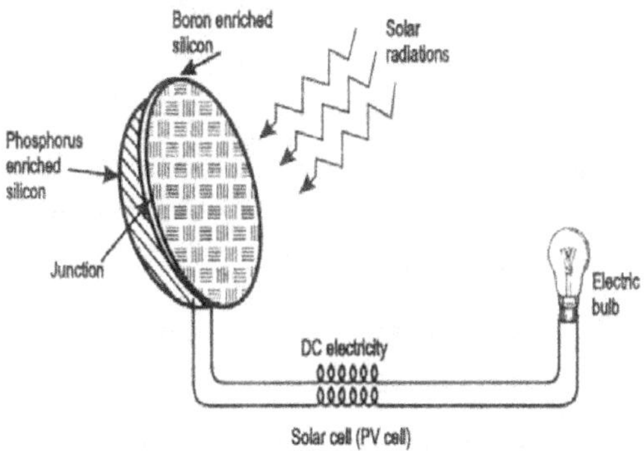

Solar Light

2. *Solar Pumps Run by Solar Battery*

When a large number of solar cells are connected in series it forms a solar battery. The solar battery produces more electricity which is enough to run, water pump, streetlights, etc. They are also used in remote areas where conventional electricity supply is a problem.

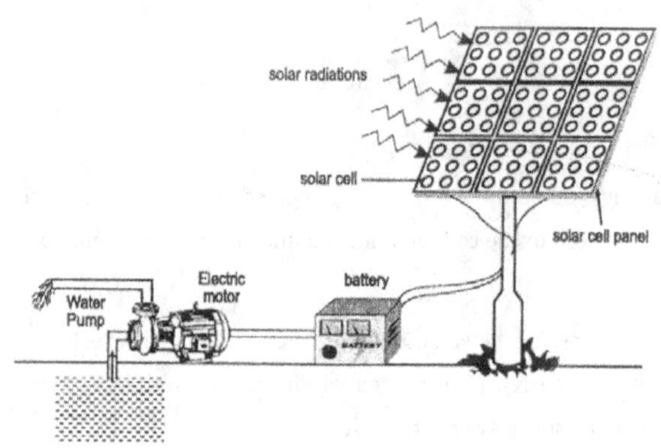

Solar Pump Run by Solar Cells (Battery)

3. Solar cells are used in calculators, electronic watches, radios and TVs.

4. Solar cells are superior to another type of cellsbecause these are non-polluting and eco-friendly.

5. Solar energy can be stored in Ni-Cd batteries and lead-acid batteries.

6. Solar cells can be used to drive vehicles.

7. Solar cells, made of silicon, are used as a source of electricity in spacecraft and satellites.

Advantages of Solar Cells

1. Solar cells can be used in remote and isolated areas, forests and hilly regions.

2. Maintenance cost is low.

3. Solar cells are noise and pollution free.

4. Their lifetime is long.

Disadvantages

1. Capital cost is higher.

2. Storage of energy is not possible.

4.7. Wind Energy

Moving air is called the wind. *Energy recovered **from the** force **of** the wind is called wind energy.*

The energy possessed by the wind is because of its high speed. The wind energy is harnessed by making use of windmills.

Methods of Harnessing Wind Energy

1. Windmills

The strike of blowing the wind on the blades of the windmill makes it rotating continuously. The rotational motion of the blade drives a number of machines like water pump, flour mills and electric generators.

Nowadays windmill uses large sized propeller blades and is connected to a generator through a shaft.

Windmills are capable of generating about 100 kW electricity.

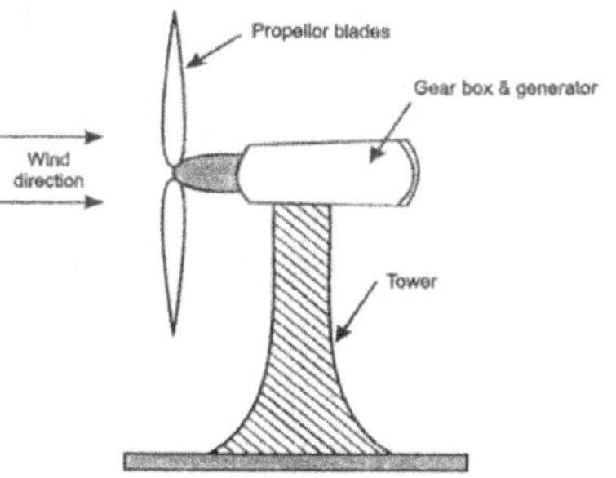

Windmill

2. Wind Farms

When a large number of windmills are installed and joined together in a definite pattern it forms a wind farm. The wind farms, produce a large amount of electricity.

Condition

The minimum speed required for the satisfactory working of a wind generator is 15 km / hour.

3. Other Methods

Other methods adopted for harnessing wind energy are

a) Skysail.

b) Ladder mill.

c) Kite ship (Large free-flying sails).

d) Sky wind power (Flying electric generator).

e) Briza technologies (Hovering wind turbine).

f) Sequoia automation (The kite wind generator).

Advantages (or) Merits of Wind Energy

a) It does not cause any air pollution.

b) It is very cheap and economic.

c) It is renewable.

d) It does not cause any pollution.

Disadvantages (or) Demerits

1. Public resists for locating the wind forms in populated areas due to noise generated by the machines and loss of aesthetic appearance.
2. Wind forms located on the migratory routes of birds will cause hazards.
3. Wind farms produce unwanted sound.
4. Wind turbines interfere with electromagnetic signals (TV, Radio signals).

4.8. Fuel Cells

Definition

A fuel cell is a voltaic cell, which converts the chemical energy of the fuels directly into electricity without combustion.It converts the energy of the fuel directly into electricity. In these cells, the reactants, products and electrolytes pass through the cell.

$$\text{Fuel+oxygen} \rightarrow \text{Oxidation products + Electricity}$$

Examples : *Hydrogen-oxygen fuel cell; Methyl alcohol-oxygen fuel*

Hydrogen-Oxygen Fuel Cell

A hydrogen-oxygen fuel cell is the simplest and most successful fuel cell, in which the fuel-hydrogen and the oxidiser-oxygen and the liquid electrolyte are continuously passing through the cell.

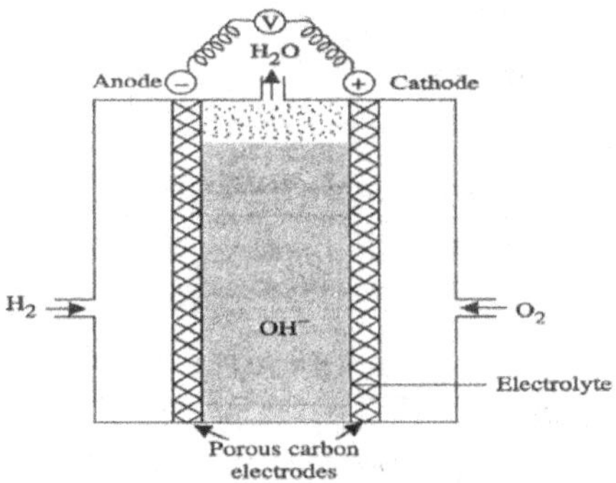

H₂-O₂ Fuel Cell

Description

It consists of two porous electrodes anode and cathode. These porous electrodes are made of compressed carbon containing a small amount of catalyst (Pt, Pd, Ag). In between the two electrodes an electrolytic solution such as 25% KOH or NaOH is filled. The two electrodes are connected to the voltmeter.

Working

Hydrogen (the fuel) is bubbled through the anode compartment, where it is oxidised. The oxygen (oxidizer) is bubbled through the cathode compartment, where it is reduced.

Various Reactions

At Anode

Hydrogen gas passed through the anode, is oxidised with the liberation of electrons which then combine with hydroxide ions to form water.

$$H_2 \longrightarrow 2H^+ + 2e^-$$

$$2H^+ + 2OH^- \longrightarrow 2H_2O$$

$$\overline{H_2 + 2OH^- \longrightarrow 2H_2O + 2e^-}$$

Multiply by 2: $\quad 2H_2 + 4OH^- \longrightarrow 4H_2O + 4e^-$

At Cathode

The electrons, produced at the anode, pass through the external wire to the cathode where it is absorbed by oxygen and water to produce hydroxide ions.

$$O_2 + 4e^- \longrightarrow 2O^{2-}$$

$$2O^{2-} + 2H_2O \longrightarrow 4OH^-$$

$$\overline{O_2 + 2H_2O + 4e^- \longrightarrow 4OH^-}$$

Overall cell reaction

At anode: $2H_2 + 4OH^- \longrightarrow 4H_2O + 4e^-$

At cathode: $O_2 + 2H_2O + 4e^- \longrightarrow 4OH^-$

$$2H_2 + O_2 \longrightarrow 2H_2O$$

➡ The emf of the cell $= 0.8$ to 1.0 V

Fuel Battery

When a large number of fuel cells are connected in series, it forms fuel battery.

Advantages of Fuel Cells

- Fuel cells are efficient (75%) and take less time for the operation.
- It is pollution free technique.
- It produces electric current directly from the reaction.
- A fuel and an oxidizer.
- It produces drinking water.

Disadvantages

1. Fuel cells can not store electric energy as other cells do.
2. Electrodes are expensive and short lived.
3. Storage and handling of hydrogen gas are dangerous.

Applications

1. H_2-O_2 fuel cells are used as an auxiliary energy source in space vehicles, submarines or other military vehicles.
2. In the case of H_2-O_2 fuel cells, the product of water is proved to be a valuable source of fresh water by the astronauts.

4.9. Batteries

Introduction

In electrochemical cells, the chemical energy is converted into electrical energy. The cell potential is related to the free energy change (AG). In an electrochemical cell, the system does work by transferring electrical energy from an electric circuit. Thus, AG for a reaction is a measure of the maximum useful work, that can be obtained from a chemical reaction.

ie., AG = maximum useful work

But we know that maximum useful work = nFE

When a cell operates, work is done in the surroundings (flow of electricity).

$$\Delta G = -nFE \quad \text{or} \quad \Delta G < 0$$

A decrease in free energy is indicated by (-)ve sign.

One of the main uses of the galvanic cells is the generation of portable electrical energy. These cells are known as batteries.

Battery

A battery is an arrangement of several electrochemical cells connected in series, that can be used as a source of direct electric current)

A Cell: It contains only one anode and cathode. A Battery: It contains several anodes and cathodes.

Requirements of a Battery

A useful battery should fulfil the following requirements

1. It should be light and compact for easy transport.
2. It should have along life,both when it is being used and when it is not used.
3. The voltage of the battery should not vary appreciably during its use.

4.10. Types of Battery

1. Primary Battery (or) Primary cells (or) Non-reversible Battery

In these cells, the electrode and the electrode reactions cannot be reversed by passing an external electrical energy. The reactions occur only once and after use they become dead. Therefore, they are not chargeable

Examples: Dry cell, mercury cell.

2. Secondary Battery (or) Secondary cells(or) Reversible Battery

In these cells, the electrode reactions can be reversed by passing an external electrical energy. Therefore, they can be recharged by passing electric current and used again and again. These are also called Storage cells (or) Accumulators.

Examples : Lead acid storage cell, Nickel-cadmium cell.

3. Flow Battery (or) Fuel Cells

In these cells, the reactants, products and electrolytes are continually passing through the cell. In this chemical energy gets converted into electrical energy. Eg.: *Hydrogen-oxygen fuel cell.*

4.11. Alkaline Battery

Description

An alkaline battery has improved the form of the dry cell, in which the electrolyte NH_4C1 is replaced by KOH. The alkalinebattery consists of a zinc cylinder filled with an electrolyte consisting of powdered Zn,KOH and MnO_2 in the form of paste using starch and water. A carbon rod (graphite), acts as a cathode, is immersed in the electrolyte in the centre of the cell. The outside cylindrical zinc body acts as an anode.

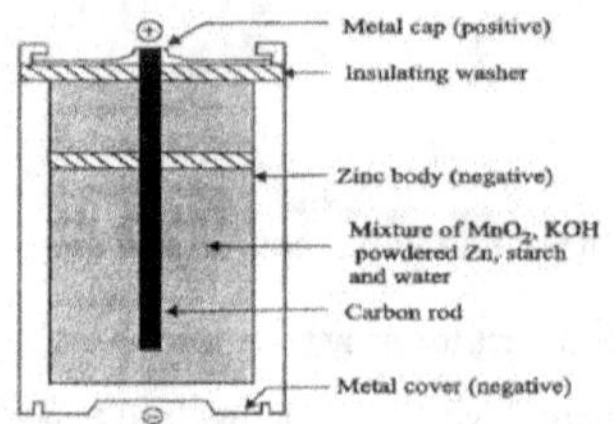

Cell Reactions

At anode: $Zn_{(s)} + 2OH^-_{(aq)} \longrightarrow Zn(OH)_{2\,(s)} + 2e^-$

At cathode:

$$2MnO_{2\,(s)} + H_2O(l) + 2e^- \longrightarrow Mn_2O_{3\,(s)} + 2OH^-_{(aq)}$$

Overall Cell Reaction

$$Zn_{(s)} + 2MnO_{2\,(s)} + H_2O_{(l)} \longrightarrow Zn(OH)_{2\,(s)} + Mn_2O_{3\,(s)}$$

In cathode reaction. Mn is reduced from +4 oxidation state to +3 oxidation state.

➡ **The emf of the cell is 1.5 V.**

Advantages of Alkaline Battery Over Dry Battery

The main advantages of the alkaline battery over dry battery are

1. Zinc does not dissolve readily in a basic medium.

2. The life of the alkaline battery is longer than the dry batterybecause there is no corrosion on Zn.

3. The alkaline battery maintains its voltage, as the current is drawn from it.

Uses

It is used in calculators, watches etc.,

4.12. Lead Storage Cell or Lead Accumulator or Acid Storage Cell

Storage Cell

A lead-acid storage cell is a secondary battery, which can operate both as a voltaic cell and as an electrolytic cell. When it acts as a voltaic cell, it supplies electrical energy and becomes "run down". When it is recharged, the cell operates as an electrolytic cell.

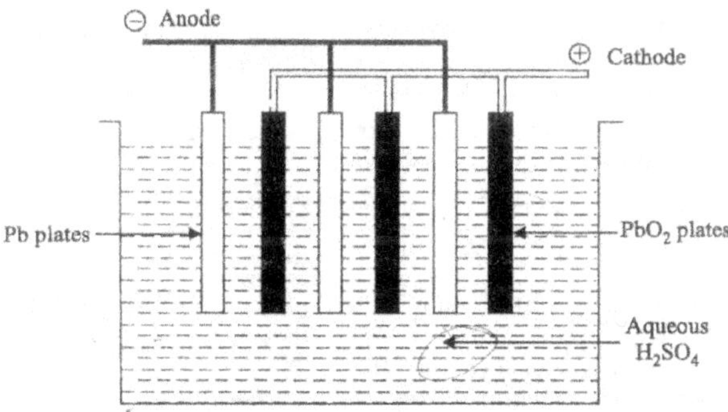

Lead Storage Cell

Description

A lead-acid storage battery consists of a number of (3 to 6) voltaic cells connected in series to get 6 to 12 V battery.

In each cell, the anode is made of lead. The cathode is made of lead dioxide PbO2_or a grid made of lead, packed with PbO2. A number of lead plates (anodes) are connected in parallel and a number of PbO2 plates (cathodes) are also connected in parallel.

Various plates are separated from the adjacent ones by insulators like rubber or glass fibre. The entire combinations are then immersed in dil. H_2SO_4 (38% by mass) having a density of 1.30 gm/ml.

The cell may be represented as,

$$Pb \mid PbSO_4 \mid\mid H2SO_4 \text{ (aq)} \mid PbO_2 \mid Pb$$

Working (Discharging)

When the lead-acid storage battery operates, the following reaction occurs.

At anode: Lead is oxidised to Pb ions, which further combines with SO_4 forms insoluble $PbSO_4$.

$$Pb_{(s)} + SO_4^{2-}{}_{(aq)} \underset{charging}{\overset{discharging}{\rightleftharpoons}} PbSO_{4\,(s)} + 2e^-$$

At cathode: PbO_2 is reduced to Pb^{2+} ions, which further combines with SO_4^{2-} forms insoluble $PbSO_4$.

$$PbO_{2\,(s)} + 4H^+ + SO_4^{2-} + 2e^- \underset{charging}{\overset{discharging}{\rightleftharpoons}} PbSO_{4\,(s)} + 2H_2O$$

Overall cell reaction during use (discharging):

$$Pb_{(s)} + PbO_{2\,(s)} + 2H_2SO_{4\,(aq)} \underset{charging}{\overset{discharging}{\rightleftharpoons}} 2PbSO_{4\,(s)}$$
$$+ 2H_2O + Energy$$

From the above cell reactions, it is clear that $PbSO_4$ is precipitated at both the electrodes and H_2SO_4 is used up.

As a result, the concentration of H_2SO_4 decreases and hence the density of H_2SO_4 falls below 1.2 gm/ml. So the battery needs recharging.

Recharging the Battery

The cell can be charged by passing electric current in the opposite direction. The electrode reaction gets reversed.

As a result, Pb is deposited on anode and PbO_2 on the cathode. The density of H_2SO_4 also increases.

The net reaction during charging is

$$2PbSO_{4(s)} + 2H_2O + Energy \underset{discharging}{\overset{charging}{\rightleftharpoons}} Pb_{(s)} + PbO_{2\,(s)}$$
$$+ 2H_2SO_{4\,(eq)}$$

Advantages of Lead-Acid Batteries

1. It is made easily.
2. It produces very high current.
3. The self-discharging rate is low when compared to other rechargeable batteries.
4. It also acts effectively at low temperature.

Disadvantages of Lead-Acid Batteries

1. Recycling of this battery causes environmental hazards.
2. Mechanical strain and normal pumpingreducebattery capacity.

Uses

1. The Lead storage cell is used to supply current mainautomobiles such as cars, buses, trucks, etc

2. It is also used in gas engine ignition, telephone exchanges, hospitals, power stations, etc.,

4.13. Nickel-Cadmium Cell (or) Nicad Battery

This is also a rechargeable battery.

Description

Nickel-cadmium cell consists of a cadmium anode and a metal grid containing a paste of NiO_2 acting as a cathode The electrolyte in this cell is KOH.

It is represented as,

$$Cd \mid Cd(OH)_2 \ 11 \ KOH_{(aq)} \mid NiO_2 \mid Ni$$

Working (Discharging)

When the Nicad battery operates, at the anode cadmium is oxidised to Cd^+ ions and insoluble $Cd(OH)_2$ is formed. It produces about 1.4V.

At anode: Cadmium is oxidised to Cd^+ and further it combines with OH^- ions to form $Cd(OH)_2$.

$$Cd_{(s)} + 2OH^- \xrightleftharpoons[\text{charging}]{\text{dicharging}} Cd(OH)_{2\,(s)} + 2e^-$$

At cathode: NiO_2 is reduced to Ni^{2+} ions which further combine with OH^- ions to form $Ni(OH)_2$.

$$NiO_{2\,(s)} + 2H_2O + 2e^- \xrightleftharpoons[\text{charging}]{\text{dicharging}} Ni(OH)_{2\,(s)} + 2OH^-$$

Overall reaction during use (discharging):

$$Cd_{(s)} + NiO_{2\,(s)} + 2H_2O \xrightleftharpoons[\text{charging}]{\text{dicharging}} Cd(OH)_{2\,(s)} + Ni(OH)_{2\,(s)}$$
$$+ \text{Energy.}$$

From the above cell reactions, it is clear that there is no formation of gaseous products, the products $Cd(OH)_2$ and $Ni(OH)_2$ adhere well to the surfaces. This can be reconverted by recharging the cell.

Recharging the Battery

The recharging process is similar to lead storage battery. When the current is passed in the opposite direction, the electrode reaction gets reversed. As a result, Cd gets deposited on anode and NiO_2 on the cathode.

The net reaction during charging is

$$Cd(OH)_{2\,(s)} + Ni(OH)_{2\,(s)} + Energy \underset{discharging}{\overset{charging}{\rightleftharpoons}} Cd_{(s)} + NiO_{2\,(s)} + 2H_2O$$

Advantage

1. It is smaller and lighter.
2. It has a longer life than lead storage cell.
3. Like a dry cell, it can be packed in a sealed container.

Disadvantage

It is more expensive than lead storage battery.

Uses

It is used in calculators, electronic flash units, transistors and cordless appliances.

4.14. Lithium Battery

Lithium battery is a solid state battery because instead of liquid or a paste electrolyte, the solid electrolyte is used.

Construction

The lithium battery consists of a lithium anode and a TiS_2 cathode. A solid electrolyte, generally a polymer, is packed in between the electrodes. The electrolyte (polymer) permits the passage of ions, but not that of electrons.

Working (Discharging)

When the anode is connected to the cathode, lithium ions move from anode to cathode. The anode is elemental lithium, which is the source of the lithium ions and electrons. The cathode is a material capable of receiving the lithium ions and electrons.

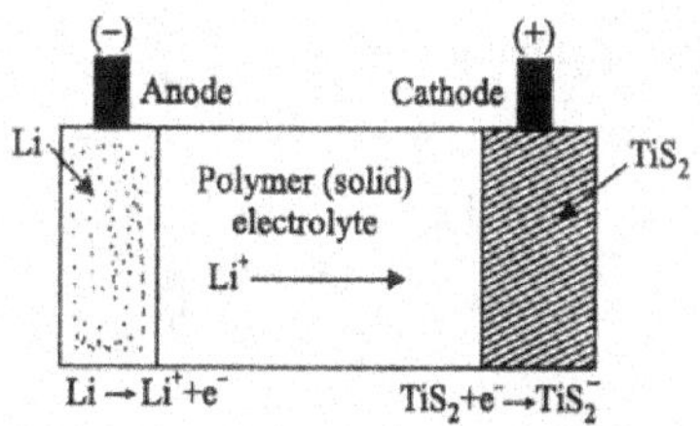

Solid State Lithium Battery

At anode: $Li_{(s)} \rightarrow Li^+ + e^-$

At cathode: $TiS_{2(s)} + e^- \rightarrow TiS_2$

Overall Reaction

$Li_{(s)} + TiS2_{(s)} \rightarrow L_{(s)}$

Recharging the Battery

The lithium battery can be recharged by supplying an external current, which drives the lithium ions back to the anode. The overall reaction is

$LiTiS_2 \rightarrow Li^+ + TiS_2^-$

This cell is rechargeable and produces a cell voltage of 3.0 V.

Other types of Secondary Lithium Batteries.

1. Li/Mnp_2
2. Li/V_2O_5
3. Li/MoO_2
4. Li/Cr_3O_8

Advantages of Li battery

It is the cell of future, why?

1. Its cell voltage is high, 3.0 V.
2. Since Li is a light-weight metal, only 7g (1 mole) material is required to produce 1 mole of electrons.
3. Since Li has the most negative E° value, it generates a higher voltage than the other types of cells.
4. Since all the constituents of the battery are solids there is no risk of leakage from the battery.
5. This battery can be made in a variety of sizes and shapes.

Disadvantages of Li-Battery

Li battery is more expensive than other batteries.

Uses

Button sized Li batteries are used in calculators, watches, cameras, mobile phones, laptop computers, etc.,

Lithium - Sulphur Battery

A Lithium-Sulphur battery is a rechargeable battery. Its anode is made of Li. Sulphur is the electron acceptor, the electron from Li is conducted to S by a graphite cathode. |3-Alumina ($NaAl_{11}O_{17}$) is used as the solid electrolyte, which separates the anode and liquid sulphur.

This solid electrolyte allows the Li^+ ions to migrate to equalise the charge, but will not allow the big polysulphide productions.

This battery is operated at high temperatures as Li and S should be in their molten states.

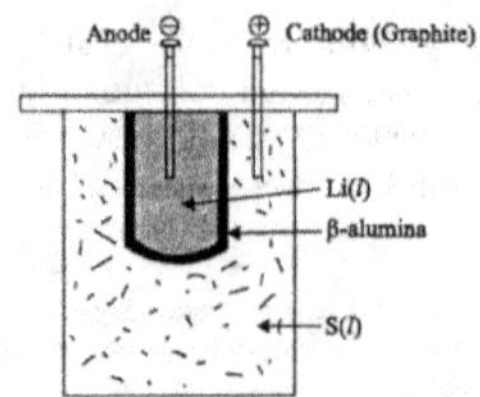

The Lithium-Sulphur Battery

Various Reactions

The various electrode reactions are

At anode: $\quad 2Li \longrightarrow 2Li^+ + 2e^-$

At cathode: $\quad S + 2e^- \longrightarrow S^{2-}$

Overall reaction: $\quad 2Li + S \longrightarrow 2Li^+ + S^{2-}$

The S^{2-} ions, formed, react with elemental sulphur to form the polysulphide ion.

$$S^{2-} + nS \longrightarrow [S_{n+1}]^{2-}$$

The direct reaction between Li and S is prevented by the alumina present in the cell.

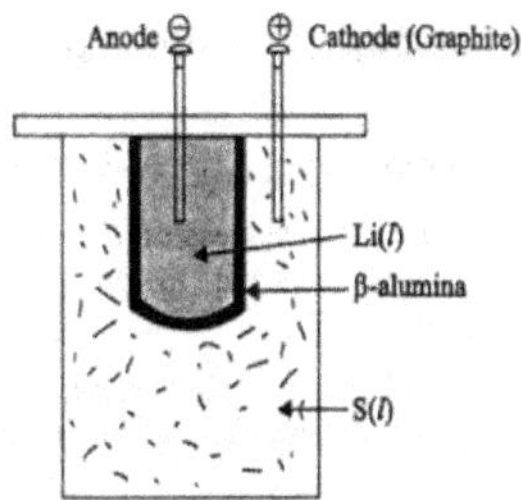

The Lithium-Sulphur Battery

VArious Reactions

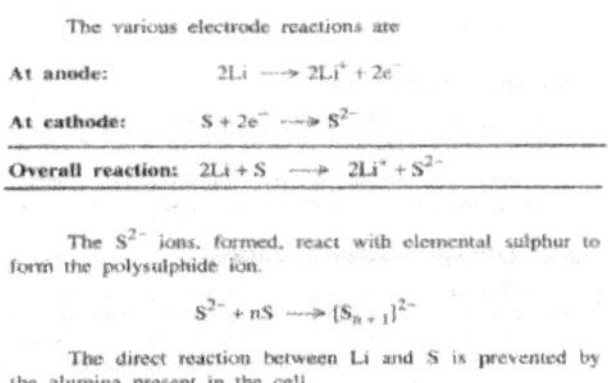

The various electrode reactions are

At anode: $2Li \longrightarrow 2Li^+ + 2e^-$

At cathode: $S + 2e^- \longrightarrow S^{2-}$

Overall reaction: $2Li + S \longrightarrow 2Li^+ + S^{2-}$

The S^{2-} ions, formed, react with elemental sulphur to form the polysulphide ion.

$$S^{2-} + nS \longrightarrow [S_{n+1}]^{2-}$$

The direct reaction between Li and S is prevented by the alumina present in the cell.

Advantages of Li-S Battery

1. Li-S battery has light weight, unlike the lead acid battery.

2. It possesses a high energy density.

3. It is used in electric cars.

UNIT V

ANALYTICAL TECHNIQUES

5.1. Introduction

Analytical technique (or) Spectroscopy is one of the most powerful tools available for the study of the atomic and molecular structure and is used in the analysis of a most of the samples. Spectroscopy deals with the study of the interaction of electromagnetic radiation with the matter. During the interaction, the energy is absorbed or emitted by the matter. The measurements of this radiation frequency (absorbed or emitted) are made using spectroscopy.

5.2. Types of Spectroscopy

The study of spectroscopy can be carried out under the following headings

1. Atomic spectroscopy.
2. Molecular spectroscopy.

1. Atomic Spectroscopy

It deals with the interaction of the electromagnetic radiation with atoms. During which the atoms absorb radiation and gets excited from the ground state electronic energy level to another.

2. Molecular Spectroscopy

It deals with the interaction of electromagnetic radiation with molecules. This results in transition between rotational, vibrational and electronic energy levels.

Differences between Molecular Spectra and Atomic Spectra

Atomic spectra	Molecular spectra
1. It occurs from the interaction of atoms + electromagnetic radiation.	It occurs from the interaction of molecules + electromagnetic radiation.
2. Atomic spectra is a line spectra.	Molecular spectra is a complicated spectra.
3. It is due to electronic transition in an element.	It is due to virbrational, rotational and electronic transition in a molecule.

5.3. Spectrum

How does a Spectrum Arise?

1. *Absorption Spectrum*

Consider a molecule having only two energy levels $E1$ and $E2$ as shown in the figure. 5.1.

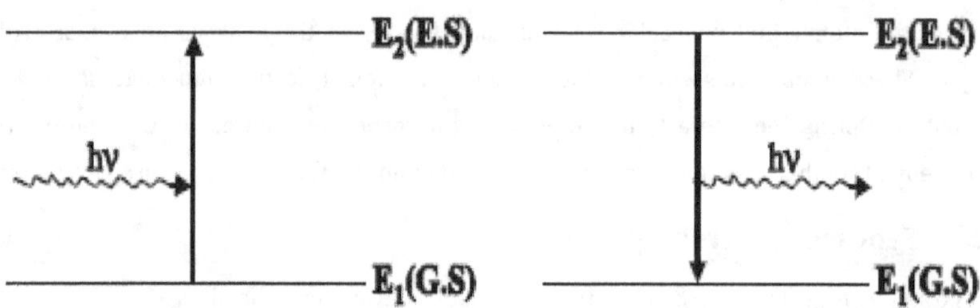

(a) Absorption Spectrum   (b) Emission Spectrum

- When a beam of electromagnetic radiation is allowed to fall on a molecule in the ground state, the molecule absorbs a photon of energy *hv and* undergoes a transition from the lower energy level to the higher energy level.

- The measurement of this decrease in the intensity of radiation is the basis of absorption spectroscopy. The spectrum, thus obtained is called the ***absorption spectrum***. (Fig 5.1. a)

2. *Emission Spectrum*

If the molecule comes down from the excited state to the ground state with the emission of photons of energy *hv*, the spectrum obtained is called ***emission spectrum***.

5.4. Photophysical Law

The absorption of light in the visible and near UV region by a solution is governed by a photophysical law known as Beer-Lambert's law.

Lambert's Law

Lambert's law states that "when a beam of monochromatic radiation is passed through a **homogeneous absorbing medium** the rate of decrease of intensity of the radiation I' with a thickness of absorbing medium '*dx*' is Proportional to the **intensity of the incident radiation** 'I'.

It is mathematically expressed as

$$\frac{-dI}{dx} = kI \qquad \text{............ (1)}$$

where. k = absorption coefficient.

On integrating the equation (1) between limits $I = I_0$ at $x = 0$ and $I = I$ at $x = x$, we get

$$\int_{I_0}^{I} \frac{dI}{I} = - \int_{0}^{x} k\,dx$$

$$\boxed{\ln \frac{I}{I_0} = - kx} \qquad \text{............ (2)}$$

The equation (2) is known as Lambert's law.

Beer's Law (or) Beer-Lambert's Law

Beer extended the above equation (2) to solutions of the compound in a transparent solvent.

According to this law, "when a beam of monochromatic radiation is passed through a *solution of an absorbing substance*, the rate of decrease of intensity of radiation 'I' with thickness of the absorbing solution 'dx' is proportional to the *intensity of incident radiation 'I'* as well as **the concentration of the solution 'C'.**"

It is mathematically represented as

$$\frac{-dI}{dx} = kIC \qquad \text{............ (3)}$$

where. k = proportionality constant.

On integrating the equation (3) between limits $I = I_0$ at $x = 0$ and $I = I$ at $x = x$, we get

$$\int_{I_0}^{I} \frac{dI}{I} = - \int_{0}^{x} kC\,dx$$

$$\ln \frac{I}{I_0} = - kCx \quad \text{(or)} \quad 2.303 \log \frac{I}{I_0} = - kCx$$

$$\text{(or)} \quad \log \frac{I_0}{I} = \frac{k}{2.303} Cx$$

$$\text{(or)} \quad \boxed{A = \varepsilon Cx} \qquad \text{............ (4)}$$

where. $\varepsilon = \dfrac{k}{2.303}$ = molar absorptivity (or) molar extinction coefficient

$$\log \frac{I_0}{I} = A = \text{Absorbance} \quad \text{(or) Optical density}$$

The equation (4) is called Beer-Lambert's law. ***Thus, the absorbance (A) is directly proportional to molar concentration (C) and thickness (or) path length (x).***

Application of Beer-Lambert's law

Determination of Unknown Concentration

The first absorbance 'A_s' of a standard solution of **known concentration** 'C_s' is measured, then according to Beer-Lambert's law.

$$A_s = \varepsilon C_s x$$

$$\frac{A_s}{C_s} = \varepsilon x \qquad \dots\dots\dots (5)$$

Now, absorbance 'A_u' of a solution of **unknown concentration** C_u is measured. Now we have

$$A_u = \varepsilon C_u x$$

$$\frac{A_u}{C_u} = \varepsilon x \qquad \dots\dots\dots (6)$$

From equation (5) and (6), we get.

$$\frac{A_s}{C_s} = \frac{A_u}{C_u}$$

$$\therefore \quad \boxed{C_u = \frac{A_u}{A_s} \times C_s} \qquad \dots\dots\dots (7)$$

since the values of A_u and A_s are experimentally determined and C_s is known. The value C_u (unknown concentration) can be calculated from the equation (7).

Limitations of Beer-Lambert's Law

1. Beer-Lambert's law is not obeyed if the radiation used is not monochromatic.
2. It is applicable only for dilute solutions.
3. The temperature of the system should not be allowed to vary to a large extent.
4. It is not applied to suspensions.
5. The deviation may occur, if the solution contains impurities.
6. Deviation also occurs if the solution undergoes polymerization (or) dissociation.

5.5. Colorimetry

Colorimetry is concerned with the visible region (400-750 nm) of the spectrum. The instrument, used for measuring the absorption of radiant energy in the visible region of the substances is called **colorimeter.**

Principle

This method is convenient for the coloured substances or coloured solutions. The intensity of colour can be easily measured by using a photoelectric colorimeter, from which the concentration of the coloured solution can be obtained by using the Beer-Lambert's law.

If the substance is colourless, then a suitable complexing agent is added to the solution so that a coloured complex is obtained, which can absorb the light.

Example

For the estimation of cuprous ions, a complexing agent, ammonium hydroxide, is added to get a blue coloured solution.

Instrumentation

A. Components

All the colorimeters have the following components.

1. *Radiation Sources*

The wavelength range of visible light lies between 400-750 nm. In this region, a tungsten-filament lamp is most widely used.

2. *Filter (or) Monochromator*

It is an instrument, which allows the light of the required wavelength to pass through, but absorbs the light of other wavelengths.

3. *Slits*

a) *Entrance Slit*

It provides a narrow source of the light.

b) *Exit Slit*

It selects a narrow band of the dispersed spectrum for observation by the detector.

4. Cell

The cell, holding the test sample (usually a solution), should be transparent. For the visibleregion, the cell is made of colour-corrected fused glass.

5. Detector

It is used for measuring the radiant energy transmitted through the sample. Photosensitive devices are used to detect radiations. These detectors produce current, which is directly proportional to the intensity of the incident radiation.

6. Meter

It is used to measure directly the fraction of light absorbed.

B. Working of Colorimeter

1. In a colorimeter, a narrow beam of light is passed from the radiation source through the test solution (cell) towards a sensitive detector (photocell).
2. Usually, acolorimeter is provided with the arrangement of filter and slits, which select the light of required wavelength.
3. The detector (photocell) generates the current, which is proportional to the amount of light transmitted by the solution.
4. The amount of light transmitted depends on the depth of colour of the test solution. Thus, the current from the

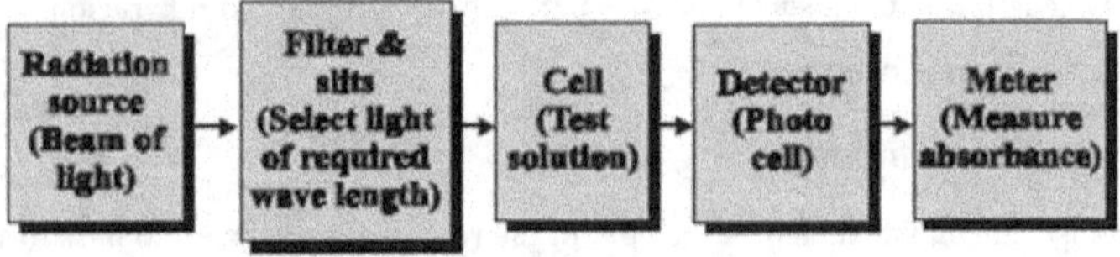

Block Diagram of Colorimeter

photocell will be more when the light transmitted is more. This will be possible only if the coloured solution is most dilute.

$$\text{Current} \propto \text{Light transmitted} \propto \frac{1}{\text{Concentration}}$$

The transmitted light is now a day allowed sending through a meter, which is calibrated to show not the fraction of light transmitted but the fraction of light absorbed. The light absorbed is proportional to the concentration of the test solution.

Estimation of Iron by Colorimetry

Principle

Fe^{3+} is a colourless solution, but Fe^{3+} forms blood red coloured complex with KCNS (or) NH_4CNS.

$$Fe^{3+} + 6\,KCNS \longrightarrow [Fe(CNS)_6]^{3-} + 6K^+$$

blood red
coloured complex

Reagents Required

(a) **Standard iron solution:** 0.865 gms of FAS is dissolved in distilled water 5-10 ml of the con. HCl is added and the solution is diluted to 1 litre. 1 ml of this solution contains 0.1 mg of Fe.

(b) **Potassium thiocyanate solution:** 20 gms of KCNS is dissolved in 100 ml of water.

(c) **1:1 con HCl:** 50 ml of the con. HCl is added to 50 ml of distilled water.

Procedure

A series of a standard solution of Fe3+ (ferric ammonium sulphate) are prepared by adding KCNS with a small amount of 1:1 con. HCl

Then the colorimeter is set at zero absorbance using a blank solution, with a proper filter. Now absorbance of each standard solution is then measured using the same filter.

A graph is plotted between absorbance vs concentration. This plot is called calibration curve and will be the straight line passing through the origin. This is according to Beer-Lambert's law.

Absorbance, $\boxed{A = \epsilon C x}$

Since, the path length (x) is fixed for a given cell, the absorbance (A) is directly proportional to concentration (C)

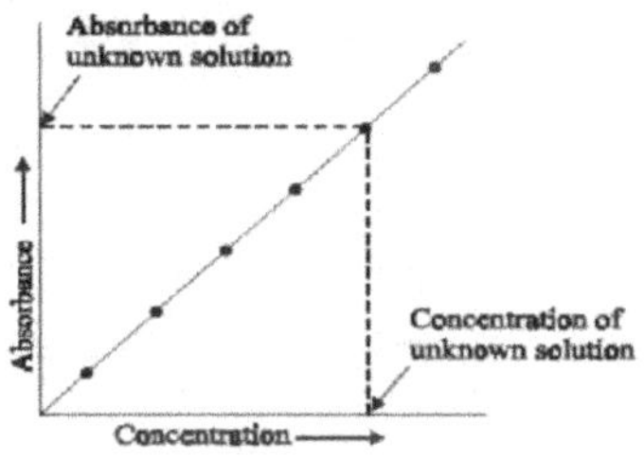

Calibraion Curve

Similarly, the absorbance of the test solution (unknown Fe^{3+} iron solution) is measured using the same colorimeter. From the calibration curve, the concentration of the unknown ferric iron solution can be evaluated.

Applications of Colorimetry

1. Molar compositions of complexes can be determined.
2. The instability constants of metal complexes are also determined.
3. Dissociation constants (Pk) of an indicator can be determined.
4. The structure of inorganic compounds, complexes (cis&trans isomer) can be determined.
5. The molecular weight of a compound can also be determined using colorimetric measurements.

5.6. Visible and Ultraviolet (UV) Spectroscopy

Principle

Visible and Ultraviolet (UV) spectra arise from the transition of valence electrons within a molecule or ion from a lower electronic energy level (ground state E0) to higher electronic energy level (excited state E1). This transition occurs due to the absorption of UV (wavelength 100-400 nm) or visible (wavelength 400-750 nm) region of the electronic spectrum by a molecule (or) ion. The actual amount of energy required depends on the difference in energy between the ground state and the excited state of the electrons.

$$E1 - Eo = h\nu.$$

Types of Electrons Involved in Organic Molecule

The energy absorbed by an organic molecule involves the transition of valence electrons. The following three types of electrons are involved in the transition.

S. No	Electrons	Examples	Energy required to excite electrons	Present in
1.	σ-electrons	Saturated long chain hydrocarbons. (Paraffins) $(CH_3-CH_2-CH_2-CH_3)$		
2.	π-electrons	Unsaturated hydrocarbons like trienes and aromatic compounds.	UV (or) visible light	Double bond and triple bonds. (unsaturated bond)
3.	n-electrons	Organic compounds containing N. O (or) halogens.	UV radiation	Unshared (or) non bonded electrons.

Example

The three types of electrons are shown in the molecule
(HCHO).

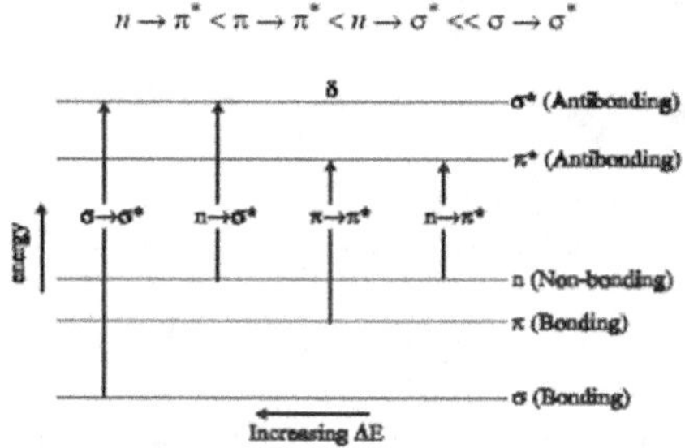

Energy Level Diagram (or) Electronic Transitions

Energy absorbed in the visible and UV region by a molecule causes transitions of valence electrons in the molecule. These transitions are

$$\sigma \to \sigma*, \; n \to \sigma*, \; n \to \pi* \; \& \; \pi \to \pi*$$

The energy level diagram for a molecule is shown in fig 8.8. The energy values for different transitions are in the following order.

$$n \to \pi^* < \pi \to \pi^* < n \to \sigma^* \ll \sigma \to \sigma^*$$

Energy LEVEL Diagream

Types of Electronic Transitions Involved in Organic Molecules

1.   **$n \to \pi*$ transitions**

$n \to \pi*$ transitions are shown by unsaturated molecules containing heteroatoms like N, O & S. It occurs due to the transition of a non-bonding lone pair of electrons to the antibonding orbitals. This transition shows a weak band and occurs in longer wavelength with low intensity.

Example

(i) *Aldehyde   &   ketone   ($CH_3 - \overset{\overset{\displaystyle \ddot{O}:}{\|}}{C} - CH_3$ &*

$CH_3 - \overset{\overset{\displaystyle H}{|}}{C} = \ddot{O}:$) (having no C = C & C ≡ C bond).

$n \to \pi^*$ transition occurs in the range of 270-300 nm.

(ii) *Aldehyde & ketone* $(CH_3 - CH = CH - \overset{\overset{\displaystyle \ddot{O}:}{\displaystyle \|}}{C} - CH_3)$ *(having double bond).*

$n \to \pi^*$ transition occurs in the range of 300-350 nm.

2. $\sigma \to \sigma^*$ transitions

$\sigma \to \sigma^*$ transitions occur in the compounds, in which all the electrons are involved in single bonds and there are no lone pair of electrons.

The energy required for $\sigma \to \sigma^*$ transition is very large. The absorption band occurs in the far UV region (120-136 nm).

Example

Saturated hydrocarbons
$$(CH_4, CH_3 - CH_3, CH_3 - CH_2 - CH_3, \text{ etc.,})$$

(i) **CH_4:** For $\sigma \to \sigma^*$: $\lambda_{max} = 121.9$ nm.

(ii) **$CH_3 - CH_3$:** For $\sigma \to \sigma^*$; $\lambda_{max} = 135$ nm.

3. $n \to \sigma^*$ transitions

$n \to \sigma^*$ transitions occur in the saturated compounds containing lone pair (non-bonding) of electrons in addition to $\sigma \to \sigma^*$ transitions. The energy required for an $n \to \sigma^*$ transition is less than that required for a $\sigma \to \sigma^*$ transition. This absorption band occurs at longer wave length in the near UV region (180-200 nm).

Example:

$$(CH_3)_3N$$

For $n \to \sigma^*$: $\lambda_{max} = 227$ nm.

and for $\sigma \to \sigma^*$: $\lambda_{max} = 99$ nm.

4. $\pi \rightarrow \pi^*$ transitions

$\pi - \pi^*$ transitions occur due to the transition of an electron from a bonding π orbital to an antibonding π^* orbital. These transition can occur in any molecule having a π electron system. Selection rule determines whether transitions to a particular π^* orbital is allowed or forbidden.

1. *UV spectrum of ethylene*

It shows *intense band* at 174 nm and *weak band* at 200 nm.Both are due to $\pi \rightarrow \pi^*$ transitions. According to selection rules, the intense band at 174 nm is due to allowed transition.

Alkyl substitution of the olefins moves the absorption to a longer wavelength. This is known as bathochromic effect or red shift. This effect increases with increase of alkyl group.

2. *UV-spectrum of unsaturated ketone*

$$\overset{\overset{\displaystyle ..}{O:}}{\underset{}{\overset{\|}{CH_2 = CH - C - CH_3}}}$$

It shows that the low intensity band at 324 nm is due to $n \rightarrow \pi^*$ transition, and high intensity band at 219 nm is due to $\pi \rightarrow \pi^*$ transition.

Important Terms Used in UV-Visible Spectrascopy

1. Chromophores (Colour producing groups)

The presence of one or more unsaturated linkages (π-electrons) in a compound is responsible for the colour of the compound, these linkages are referred to as chromophores.

Example

$$\underset{}{>}C = C< ; \ -C \equiv C- ; \ -C \equiv N; \ -N = N-; \ >C = O; \ etc.,$$

Chromophores undergo $\pi \rightarrow \pi^*$ transitions in the short wavelength regions of UV-radiations.

2. Auxochrome (Colour intensifying groups)

It refers to an atom or a group of atoms which does not give rise to absorption band on its own, but when conjugate to chromophore will cause a red shift.

Example

$- OH, - NH_2, - Cl, - Br, - I, etc.,$

Some Important Definitions Related to Change in Wavelength and Intensity

1	Bathochromic shift (Redshift)	Shift to higher wavelength (lower frequencies)
2	Hypsochromic shift (Blueshift)	Shift to lower wavelength (Higher frequencies)
3	Hyperchromic Effect	An increase in intensity
4	Hypochromic Effect	A decrease in intensity

Illustration

In chloroethylene, $CH_2=CHCl$.

$C = C$ is a chromophore.

Cl is an autochrome.

Substitution of a hydrogen atom in ethylene by a halogen atom causes a bathochromic shift and a hyperchromic effect.

Si No	Chromophore	Auxochrome
1	This group is responsible for the colour of the compound	It does not impart colour , but when conjugate to chromophore produces colour.
2.	It does not form salt	But it forms salt
3	It contains at least one multiple bonds Example: $-NO_2$, $-NO$, $-N=N-$	It contains lone pair of electrons Example: $-OH$, $-NH_2$, $-NR_2$

5.7. Instrumentation

A. Components

The various components of a visible UV spectrometer are as follows.

1. Radiations Source

Invisible – UV spectrometers, the most commonly used radiation sources are hydrogen (or) deuterium lamps.

2. Monochromators

The monochromator is used to disperse the radiation according to the wavelength. The essential elements of a monochromator are an entrance slit, a dispersing element and an exit slit. The dispersing element may be a prism or grating (or) a filter.

3. Cells (Sample Cell and Reference Cell)

The cells, containing samples or reference for analysis, should fulfil the following conditions.

- They must be uniform in construction.
- The material of construction should be inert to solvents.
- They must transmit the light of the wavelength used.

4. Detectors

There are three common types of detectors used invisible UV spectrophotometers. They are Barrier layer cell, Photomultiplier tube, Photocell.

The detector converts the radiation, falling on which, into the current. The current is directly proportional to the concentration of the solution.

5. Recording System

The signal from the detector is finally received by the recording system. The recording is done by recorder pen.

B. Working of Visible and UV Spectrophotometer

The radiation from the source is allowed to pass through the monochromator unit. The monochromator allows a narrow range of wavelength to pass through an exit slit. The beam of radiation coming out of the monochromator is split into two equal beams.

One-half of the beam (the sample beam) is directed to pass through a transparent cell containing a solution of the compound to be analysed.

The another half (the reference beam) is directed to pass through an identical cell that contains only the solvent. The instrument is designed in such a way that it can compare the intensities of the two beams.

If the compound absorbs light at a particular wavelength, then the intensity of the sample beam (I) will be less than that of the reference beam (Io).The instrument gives output graph, which is a plot of wavelength Vs absorbance of the light. This graph is known as an absorption spectrum.

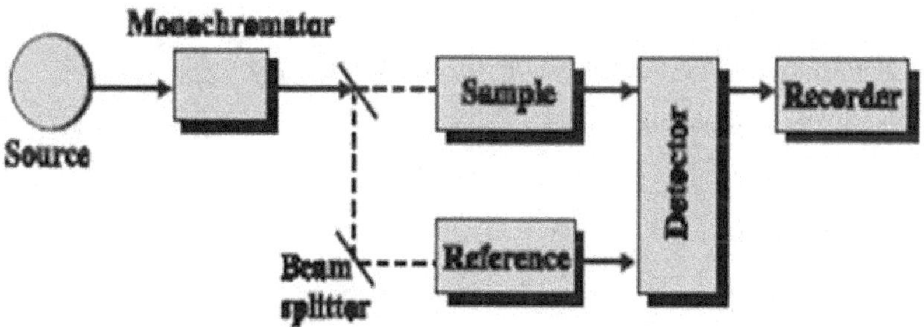

Black Diagram of visible uv Spectrophotometer

Applications

1. Predicting Relationship between Different Groups

UV spectroscopy is not useful in the detection of individual functional groups, but it is used in predicting the relationship between different groups.

i.e.,

1. between two or more C–C multiple bonds(= (or) $\equiv$ bonds).

2. between C – C & C – O double bonds.

3. between C – C double bonds and the aromatic benzene ring.

Thus, the structure of several vitamins and steric hindrance of the molecule can be determined using UV spectroscopy.

2. Qualitative Analysis

UV absorption spectroscopy is used for characterising and identification of aromatic compounds and conjugated olefins by comparing the UV absorption spectrum of the sample with the same of known compounds available in reference books.

3. Detection of Impurities

UV absorption spectroscopy is the best method for detecting impurities in organic compounds, because

1. The bands due to impurities are very intense.

2. Saturated compounds have little absorption band and unsaturated compounds have strong absorption band.

4. Quantitative analysisdetermination of Substances

UV absorption spectroscopy is used for the quantitative determination of compounds, which absorbs UV light. This determination is based on Beer's law.

5. Determination of Molecular Weight

The molecular weight of a compound can be determined if it can be converted into a suitable derivative, which gives an absorption band.

6. Dissociation constants of Acids and Bases

The dissociation constant (Pk_a) of an acid (HA) can be determined by determining the ratio of $\dfrac{[HA]}{[A]}$ spectrophotometrically from the graph plotted between absorbance vs wavelength at different pH values. This values are substituted in the equation

$$Pk_a = pH + \log \frac{[HA]}{[A]}$$

7. Study of tautomeric equilibrium

The percentage of various keto and enol forms present in a tautomeric equilibrium can be determined by measuring the strength of the respective absorption bands using UV spectroscopy.

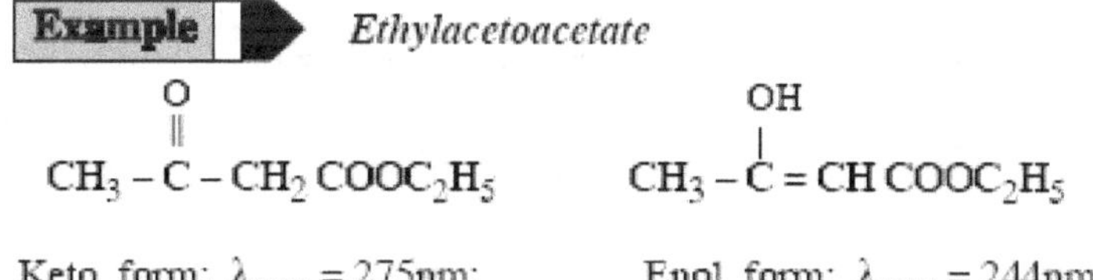

8. Studying kinetics of chemical reactions

Kinetics of chemical reactions can be studied using UV spectroscopy by following the change in concentration of a product or a reactant with time during the reaction.

9. Determination of Calcium in Blood Serum

Calcium in the blood can be determined by converting the 'Ca' present in 1 ml of the serum as its oxalate and redissolving it in H2SO4 and treating it with dilute ceric sulphate solution. The absorption of the solution is measured at 315 nm. Thus, the amount of 'Ca' in the blood serum can be calculated.

5.8. Infrared Spectroscopy

Principle

IR spectra are produced by the absorption of energy by a molecule in the infrared region and the transitions occur between vibrational levels. So, *IR* spectroscopy is also known as vibrational spectroscopy.

Range of Infrared Radiation

The range of the electromagnetic spectrum extending from 12500 to 50 cm− 1 (0.8 to 200 μ) is commonly referred to as the infrared. This region is further divided into three sub-regions.

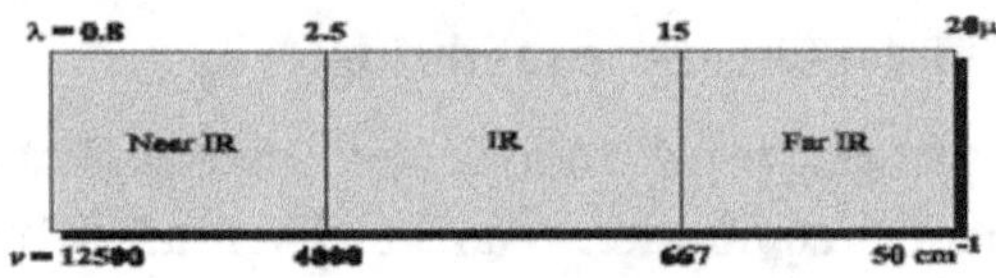

Range of IR radiaton

(i) *Near infrared:* The region is from 12500 to 4000 cm^{-1}

(ii) *Infrared (or) ordinary IR:* The region is from 4000 to 667 cm^{-1}

(iii) *Far infrared:* The region is from 667 to 50 cm^{-1}

Source of IR: Electrically heated rod of rare - earth oxides.

Molecular Vibrations and Origin of IRSpectrum

Since atoms in a molecule are continuously vibrating, molecules are also vibrating. There are two kinds of fundamental vibrations in the molecule.

1. **Stretching vibrations:** During stretching the distance between two atoms decreases or increases, but bond angle remains unaltered.

2. **Bending (or) deformation vibrations:** During bending bond angle increases and decreases but bond distance remains unaltered.

Vibrational changes depend on the masses of the atoms and their spatial arrangement in the molecule.

Finger Print Region

The vibrational spectral (*IR* spectra) region at**1400 – 700 cm– 1** gives very rich and intense absorption bands. This region is termed as **fingerprint region**. The region **4000 – 1430 cm– 1** is known as a **Group frequency region**.

Uses of Finger Print Region

1. IR spectra are often characterised as molecular fingerprints, which detect the presence of functional groups.
2. Fingerprint region is also used to identify and characterise the molecular just as a fingerprint can be used to identify a person.

Types of Stretching and Bending Vibrations

The number of fundamental (or) normal vibrational modes of a molecule can be calculated as follows.

1. For Non-Linear Molecule

A non-linear molecule containing '*n*' atoms has(3*n* – 6) fundamental vibrational modes.

Example

(i) $CH_4 \longrightarrow (3 \times 5 - 6) = 9$ *fundamental vibrational modes.*

(ii) $C_6H_6 \rightarrow (3 \times 12 - 6) = 30$ *fundamental vibrational modes.*

2. **For linear molecules**

A linear molecule containing '*n*' atoms has $(3n - 5)$ fundamental vibrational modes.

Example

$CO_2 \rightarrow (3 \times 3 - 5) = 4$ *fundamental vibrational modes.*

Illustrations

1. **Water**

Water is a bend (non-linear) triatomic molecule, and has $3n - 6$ $(3 \times 3 - 6) = 3$ fundamental vibrational modes. These modes and their corresponding frequencies are shown below (Fig. 8.11).

(i)

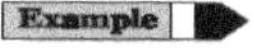

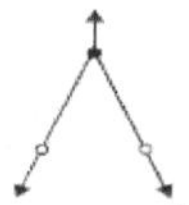

Symmetric stretching vibration

$v_1 = 3652$ cm^{-1}

(ii)

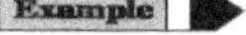

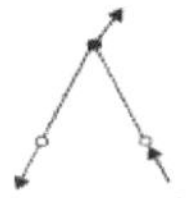

Asymmetric stretching vibration

$v_2 = 3756$ cm^{-1}

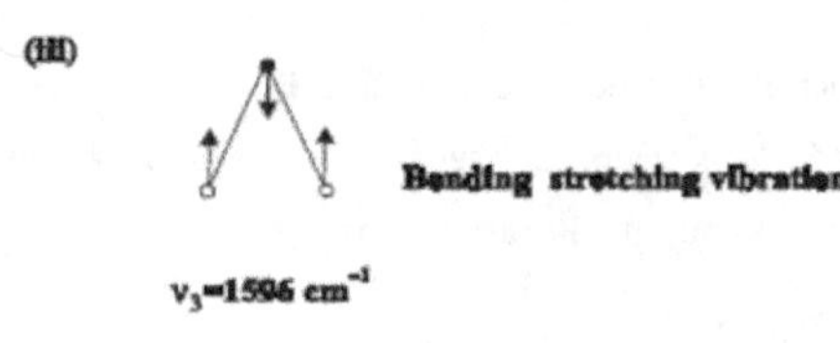

Vibrational Modes of H$_2$O

In general stretching frequencies are much higher than bending frequencies. This is because, more energy is required to stretch a bond than to bend it. All the above 3 vibrations are said to be active (change in dipolemoment during the

vibration) in the *IR* region and the *IR* spectrum of water exhibits 3 absorption bands at 1596, 3652 and 3756 cm– 1 corresponding to the bending, symmetric stretching and the asymmetric stretching vibrations respectively.

Thus, for a vibration to be active on IR, the dipole moment of the molecule must change.

2. Carbon Dioxide

Carbon dioxide is a linear triatomic molecule and has $3n - 5$ $(3 \times 3 - 5) = 4$ fundamental vibrational modes. These modes and corresponding frequencies are shown in Fig. 8.12. the four normal modes of vibration of CO2, only the asymmetric stretching and bending vibrations ie., (ii), (iii) and (iv) involve a change in dipole moment (all are *IR* - active).

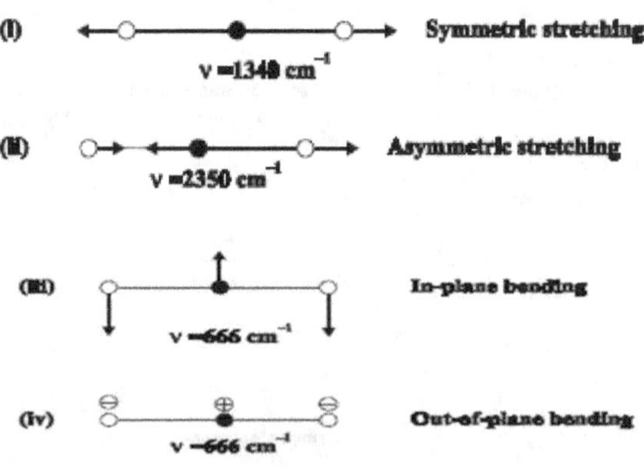

Vibrational Modes of CO$_2$

But the symmetric stretching vibration ie., (i) does not involve any change in the dipolemoment (*IR* inactive).

(Note: + and − signs indicate the motion of the corresponding atom, above and below the plane of the paper respectively).

Thus, though there are three active vibrations, two of them ((iii) and (iv)) have the same frequency, so the *IR* spectrum of CO_2 exhibits only two bands ie., one at 666 cm− 1 and another at 2350 cm− 1.

Instrumentation

A. Components

1. Radiation Source

The main source of IR radiation.

i) Nichrome wire

ii) Nernst glower, which is a filament containing oxides of Zr, Th, Ce, held together with a binder.

iii) When they are heated electrically at 1200 to 2000°C, they glow and produce IR radiation.

2. Monochromator

It allows the light of the required wavelength to pass through but absorbs the light of another wavelength.

3. Sample Cell

The cell, holding the test sample, must be transparent to IR radiation.

4. Detector

IR detectors generally convert thermal radiant energy into electrical energy. There are so many detectors, of which the followings are important.

1. Photoconductivity cell.
2. Thermocouple.
3. Pyroelectric detectors.

5. Recorder

The recorder records the signal coming out from the detector.

B. *Working of IR Spectrophotometer*

The radiation emitted by the source is split into two identical beams having equal intensity. One of the beams passes through the sample and the other through the reference sample.

When the sample cell contains the sample, the half-beam travelling through it becomes less intense.

When the two half beams (one coming from the reference and the other from the sample) recombine, they produce an oscillating signal, which is measured by the detector. The signal from the detector is passed to the recording unit and recorded.

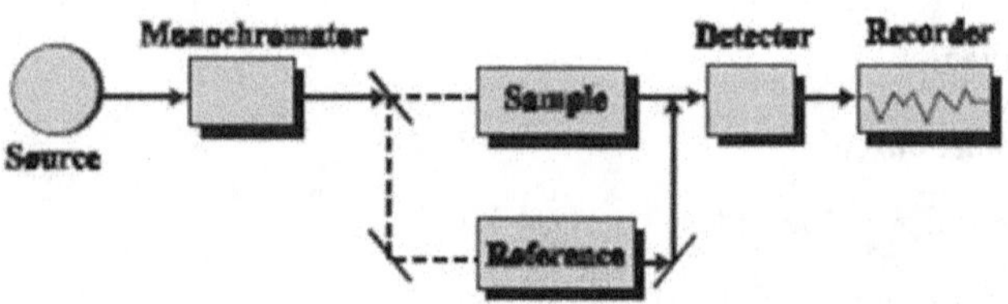

Block Diagram of Double Beam IR Spctrophotometer

Applications of IR spectroscopy

1. *Identity of the Compound can be Established*

The *IR* spectrum of the compound is compared with that of known compounds. From the resemblance of the two spectra, the nature of the compound can be established.

This is because a particular group of atoms gives a characteristic absorption band in the *IR* spectrum.

IR spectra of both benzaldehyde $(C_6H_5 - \overset{\overset{\text{H}}{|}}{C} = O)$ and phenylmethylketone $(C_6H_5 - \overset{\overset{O}{\|}}{C} - CH_3)$ show a sharp absorption peak at 1700 cm^{-1}. This indicates the presence of $C = O$ group in both the compounds.

2. Detection of functional groups

In a given environment, a certain functional group will absorb *IR* energy of very nearly the same wavelength in all molecules.

Examples

(i) Acetone $(CH_3 - \overset{\overset{\textstyle O}{||}}{C} - CH_3)$ and diethylketone $(C_2H_5 - \overset{\overset{\textstyle O}{||}}{C} - C_2H_5)$ give absorption peak at the **same place**.

(ii) But, acetic acid (CH_3COOH) and cyclobutanone

$$\begin{array}{ccc} CH_2 & - & CH_2 \\ | & & | \\ | & & C=O \\ | & & | \\ CH_2 & - & CH_2 \end{array}$$

give absorption peak at **different places**.

3. Testing purity of a sample

Pure sample will give a sharp and well - resolved absorption bands. But impure sample will give a broad and poorly resolved absorption bands. Thus by comparison with

IR spectra of pure compound, presence of impurity can be detected.

4. Study of Progress of a Chemical Reaction

The progress of a chemical reaction can be easily followed by examining the *IR* spectrum of test solution at different time intervals.

Example

The progress of oxidation of secondary alcohol to ketone is studied by getting IR spectra of test solution at different time intervals.

The secondary alcohol absorbs at 2.8 μ (~ 3570 cm− 1) due to O – H stretching. As the reaction proceeds this band slowly disappears and a new band near 5.8 μ (~ 1725 cm− 1), due to C = O stretching appears.

5. Determination of Shape (or) Symmetry of a Molecule

Whether the molecule is linear (or) non-linear (bend molecule) can be found out by *IR* spectra.

Example

IR spectra of NO2 gives three peaks at 750, 1323, and 1616 cm– 1. According to the following calculations,

1. For non-linear molecule = (3n – 6) = 3 peaks.
2. For linear molecule = (3n – 5) = 4 peaks.
3. Since the spectra show only 3 peaks, it is confirmed that NO2 molecule is a non-linear (bend) molecule.

6. To Study Tautomerism

Tautomeric equilibrium can be studied with the help of *IR* spectroscopy.

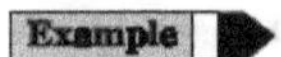

> The common systems such as keto-enol, lacto - lactum, and mercapto - thioamide , contain a group like $C = O$, $-OH$, $-NH$ (or) $C = S$. These groups show a characteristic absorption band in the IR spectrum, which enable us to find at which form predominates in the equilibrium.

7. Industrial applications

(a) Determination of structure of chemical products

During the polymerisation, the bulk polymer structure, can be determined using *IR* spectra.

(b) Determination of Molecular Weight

Molecular weight, of a compound, can be determined by measuring end group concentrations, using *IR* spectroscopy.

(c) Crystallinity

The physical structure like crystallinity can be studied through changes in *IR* spectra.

5.9. Flame Photometry (or) Flame Emission Spectroscopy

Flame photometry is a method in which, the intensity of the emitted light is measured when a atomised metal is introduced into a flame. The **wavelength** of the colour tells us **what the element is,** and the **intensity** of the colour tells us **How much of the element is present.**

Theory (or) Principle

When a metallic salt solution is introduced into a flame, the following processes will occur.

a.   The solvent is evaporated, leaving behind the solid salt particle.

b.   The salt is vaporised into the gaseous state and dissociated into atoms.

c.   Some of the atoms from the ground state are excited to a higher energy state by absorbing thermal energy from the flame.

The excited atoms, which are unstable, quickly emit photons of different wavelengths and return to the lower energy state. Then the emitted radiation is passed through the filter, which permits the characteristic wavelength of the metal under examination. It is then passed into the detector, and finally into the recorder.

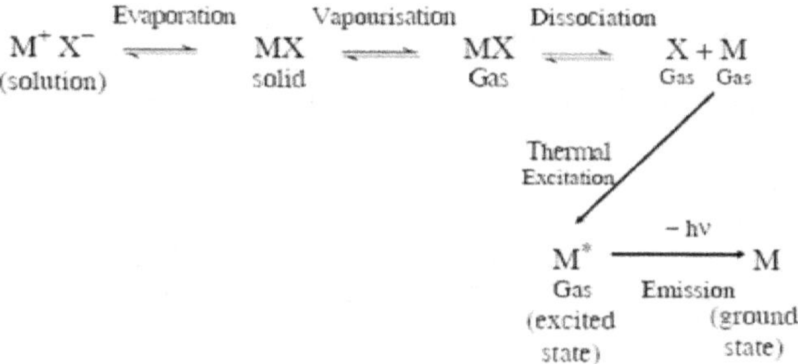

Instrumentation

A.   Components

The various components of the flame photometer are described as follows.

1.   Burner

The flame must possess the following characteristics.

1.   It should evaporate the solvent from the sample solution.

2.   It should decompose the solid into atoms.

3.   It should excite the atoms and cause them to emit radiant energy.

2.   Mirror

The radiation from the flame is emitted in all directions in space. In order to increase the amount of radiation reaching the detector, a convex mirror is used which is set behind the burner.

3. Slits

Entrance slits: It is kept between the flame and monochromator. It permits only the radiation coming from the flame and a mirror.

Exit slit: It is kept between the monochromator and detector. It prevents the entry of interfering lines.

4. Monochromator (or) Prism (or) Grating (or) Filter

It allows the light of the required wavelength to pass through but absorbs the light of other wavelengths.

5. Detector

The radiation coming out from the filter is allowed to fall on the detector, which measures the intensity of the radiation falling on it.

Photomultiplier (or) photocell is used as a detector, which converts the radiation into an electrical current.

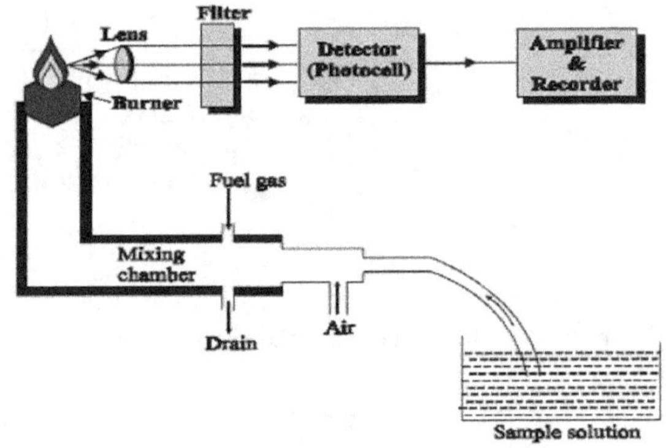

Layout of a Simple Flame Photometer

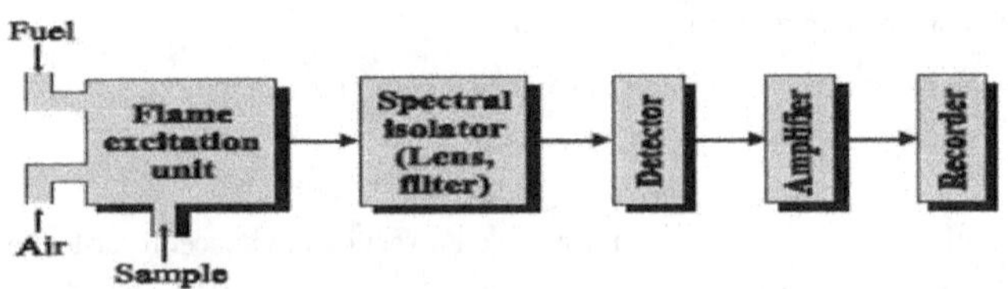

Block Diagram of a Flame Photometer

6. Amplifier & Recorder

The current coming out of the detector is weak, so it is amplified and recorded.

B. Working of Flame Photometer

Air, at a given pressure, is passed into an atomiser. The suction so-produced draws some solution of the sample into the atomiser.

Air + sample solution is then mixed with fuel gas in the mixing chamber. The Air + sample solution + fuel gas mixture is then burnt in the burner.

The radiation, emitted by burner flame, is passed successively through the lens, filter, detector, amplifier and finally into a recorder.

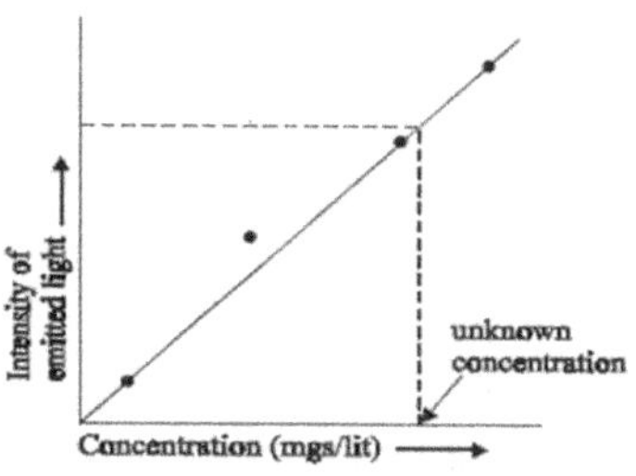

Calibration Curve

The above experiment is first carried out using a series of the standard solution, and the reading for each solution is noted. Now the graph, called calibration curve, is drawn between concentration vs intensity of emitted light (or) photometric reading.

Now the test solution (unknown) is taken and similar experiment is carried out. From the graph, the concentration of the unknown sample can be determined.

Applications of Flame Photometry

1. Estimation of Sodium by Flame Photometry

1. The instrument is switched on. Air supply and gas supply are regulated. First distilled water is sent and ignition is started.
2. After the instrument is warmed up for 10 min, the instrument is adjusted for zero reading in the display.
3. Since sodium produces a characteristic yellow emission at 589 nm, the instrument is set at $\lambda = 589$ nm and the readings are noted.

A series of standard NaCl solution (1, 2, 3, 4, 5 ... 10 ppm) is prepared and is sent one by one and the readings (intensity of emitted light) are noted. The calibration graph is drawn between the concentration Vs intensity of the emitted light. A straight line is obtained.

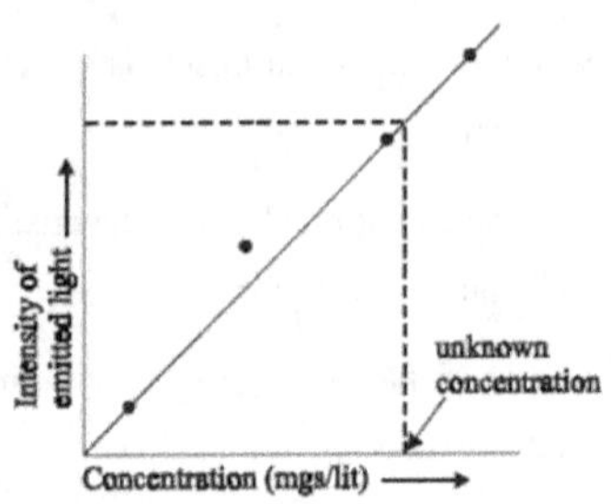

Calibration Curve

Now the unknown sodium solution is sent and the reading (intensity of the emitted light) is noted. Then the concentration of sodium in the water sample is determined from the calibration curve.

2. Qualitative Analysis

(a) The elements of group I & II. (K, Na, Li, Ca, Mg, etc) can be detected visually from the colour of the flame.

(b) Non-radiating elements such as carbon, hydrogen and halides cannot be detected using this method.

3. Quantitative Analysis

(a) The amount of the elements in a group I & II(alkali & alkaline-earth metals) can be determined from the sample.

(b) Certain transition elements, such as Cu, Fe &Mn can also be determined using flame photometry.

Example

Element	λ_{max}	Colour of flame
Ca	422 nm	Brick red
K	766 nm	Red
Na	586 nm	Yellow
Li	670 nm	Scarlet red

4. Other Applications

1. The measurement of these elements is very useful in medicine, agriculture and plant science.
2. Flame photometry is extensively used in the analysis of biological fluids and tissues.
3. In soil analysis, the elements like Na, K, Al, Ca, Fe, etc., are determined.
4. Industrial and natural waters, petroleum products, cement, glass, and metallurgical products can also be analysed by this method.

Limitations of Flame Photometry

1. It cannot be used for the determination of all metal atoms and inert gases.
2. Only liquid samples must be used.
3. It does not provide information about the molecular form of the metal present in the original sample.

5.10. Atomic Absorption Spectroscopy

Principle

Atomic absorption spectroscopy is based on the atomization of the sample followed by absorbing characteristic radiation by the ground state gaseous atoms.

When the light of the required wavelength is allowed to pass through a flame, having atoms of the metallic species, part of that light will be absorbed and the absorption will be proportional to the concentration of the atom in the flame.

Thus, in atomic absorption spectroscopy, the amount of light absorbed is determined.

Instrumentation

A. Various Components

1. Radiation Source

The radiation source should emit stable, intense,characteristic radiation of the element to be determined. The hollow cathode lamp, which consists of a glass tube containing noble gases like an argon (anode) and hollow cathode, made of the analyte metal, is generally used.

2. Chopper

A rotating wheel is interposed between the hollow cathode lamp and the flame. It breaks the steady light, from the lamp, into a pulsating light (because the recorder will record only the pulsating (alternating) current).

3. Burner (or) Flame

The flame is used for converting the liquid sample into the gaseous state. It converts the molecule into atomic vapour. Two types of burners are used

1. Total consumption burner.
2. Premixed burner.

4. Nebulisation of the Liquid Sample

Before the liquid sample enters the burner, it is first of all converted into small droplets. This method of formation of small droplets from the liquid sample is called **nebulisation**.

5. Monochromators

The monochromators select a given absorbing line from the spectral lines emitted from the hollow cathode. The most common monochromators are

1. Prisms.
2. Gratings.

6. Detectors

The photomultiplier tube is a most suitable detector. When the photon strikes the photomultiplier tube, an electric current (emf) is produced.

7. Amplifier

The electric current, from the photomultiplier detector, is fed into the amplifier, which amplifies the electric current many times.

8. ReadOut Device (or) Recorder

The signal coming out from the amplifier is recorded using a chart recorder (or) digital read-out devices.

B. Working of Atomic absorption Spectrophotometer

The characteristic radiation, obtained from the hollow cathode lamp, is passed through a flame in which the sample is aspirated.

The metallic compounds are decomposed into atoms of the element to be measured.

The atoms absorb a fraction of radiation in the flame. The unabsorbed radiation from the flame is allowed to pass through a monochromator.

From the monochromator, the unabsorbed radiation is led into the detector. From the detector, the output is amplified and measured on a recorder.

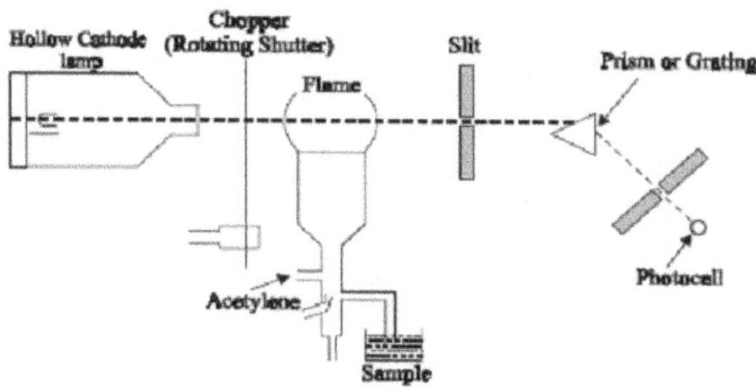

Atomic Absorption Specrtophotometer

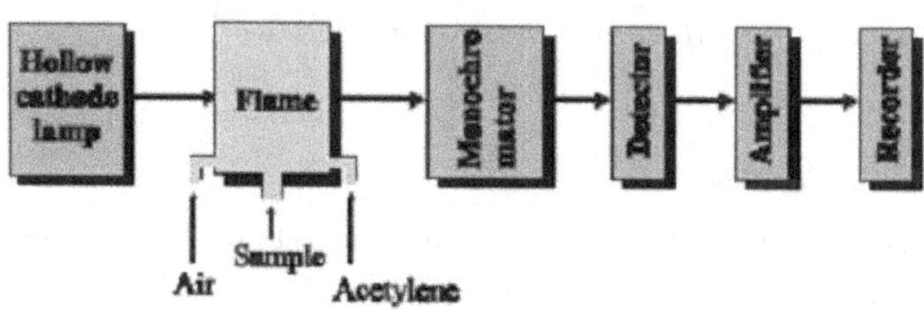

Block Diagram of Automic Absorption Spectroscopy

Quantitative Estimation of Nickel

Determination of Ni involves the following steps.

Operating parameters

Wavelength	232.0 nm
Light source	Hollow cathode lamp (Ni)
Flame type	Nitrous oxide-acetylene flame reducing (rich, red)

Step I

Preparation of standard stock Ni solution

Standard nickel solution of 1000 ppm. is prepared by dissolving 1.0 gm of nickel nitrate ($Ni(NO_3)_2$) (from Ni-metal) in minimum dil (1:1) HNO_3 and diluted to 1 litre with 1% HNO_3.

Standard calibration curve

From the standard stock solution of Ni-metal, a series of standard solutions are prepared by appropriate dilutions.

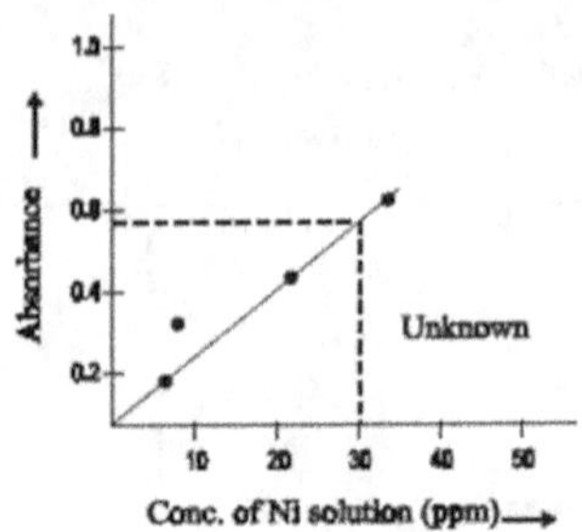

Calibrarion Curve

Now the standard solutions are aspirated one by one into the flame and absorbance are measured.

Now the graph is plotted between absorbance and concentration.

Step III

Determination of Unknown Concentration

The unknown solution of Ni is aspirated into the flame and absorbance is measured under the same condition as performed for the standard calibration.

From the absorbance, the concentration of the unknown sample is determined.

Applications of AAS

1. AAS is a rapid and accurate method, using which nearly 70-80 metals can be determined from a sample.
2. Vanadium in lubricating oils can be determined.
3. Trace elements like lead in alloys can be estimated.
4. This technique is also used in steel, metal,
5. Pharmaceutical and food industry.

Limitations of AAS

1. Preparation of sample solution is tedious and time-consuming.
2. It does not provide any information on the chemical form of the metal.
3. This technique is limited to only metals.

Question Bank for Engineering Chemistry (2016-17)

Unit I-Chemistry in Everyday Life

Part-A

1) What are Analgesic? What are the types of analgesics?

An analgesic (also known as a painkiller) is any member of the group of drugs used to relieve pain.

   a) Narcotic analgesics.

   Natural analgesics (eg. Morphine, codeine etc.)

   Synthetic analgesics (eg. pethidine, methadone etc.)

   b) Non-narcotic analgesics.

2) What are antiseptics?Give examples.

Antiseptics are antimicrobial substances that are applied to living tissue or skin to reduce the possibility of infection.

Some antiseptics are true germicides, capable of destroying microbes (bacteriocidal), whilst others are bacteriostatic and only prevent or inhibit their growth.

Egs., Alcohol, Boric acid, Iodine.

3) What is antacid? Give examples

An antacid is a substance, generally a base or basic salt, which neutralizes stomach acidity. Examples are:

   $NaHCO_3$ and/or $KHCO_3$

   $Al(OH)_3$ and $Mg(OH)_2$

   $CaCO_3$ $MgCO_3$

4) What are disinfectants? Give its types.

Disinfectants are substances that are applied to non-living objects to destroy microorganisms that are living on the objects. Air disinfectants, Aldehydes, Alcohols, Oxidizing agents and Phenolics.

5) What are artificial sweetening agents? Give examples.

   •   Artificial Sweetener, any synthetically produced, intensely sweet substance developed for use in reduced-calorie or diet foods and beverages. Egs are,

   •   Saccharin.

- Aspartame.
- Sucralose.
- Cyclamates.

6) What are sugar substitutes? Give examples.

A sugar substitute is a food additive that duplicates the effect of sugar in taste, usually with less food energy. Egs are sorbitol and Xylitol.

7) What is saccharine?

Saccharin is a synthetic organic compound that is 200-100 times sweeter as cane sugar. The sodium or calcium salt of saccharin is widely used as a diet sweetener.

8) What is sucralose? What is its use.

Sucralose is a chlorinated sugar that is about 600 times as sweet as sugar. It is produced from sucrose when three chlorine atoms replace three hydroxyl groups.

It is used in beverages, frozen desserts, chewing gum, baked goods, and other foods.

9) What are cyclamates? What is its use.

Three similar compounds, namely sodium cyclamate, calcium cyclamate, and cyclamic acid, are collectively referred to as the "cyclamates". They are about thirty times sweeter than sugar and are chemically more stable than saccharin or aspartame. Some of a cyclamate dose is excreted by the body unchanged, but some are converted to a cyclohexylamine.

10) What are non-sugar sweeteners? What is its use.

Some non-sugar sweeteners are polyols, also known as "sugar alcohols." These are, in general, less sweet than sucrose and used in a wide range of food products.

- To assist in weight loss.
- Dental care: Sugar substitutes are tooth-friendly, as they are not fermentedby the micro flora of the dental plaque.

11) What are food preservatives? Give examples.

Artificial preservatives are a group of chemical substances added to food, sprayed on the outside of food, or added to certain medications to retard spoilage, discoloration, or contamination by bacteria and other disease organisms

- Benzoates(sodium benzoate, benzoic acid).
- Nitrites (sodium nitrites), Sulphities.
- Sorbates (sodium sorbates, potassium sorbate).

12) What is BOD?

BOD is defined as "The amount of oxygen consumed by the microorganisms during the bio-chemical degradation of organic matter under aerobic conditions at 20°C for five days".

13) What is COD?

COD is defined as the amount of oxygen required for the oxidation of organic matter as well as oxidisible inorganic matter. COD is expressed as mg/L.

14) Define hard water and soft water.

The water sample, which gives a ready lather with soap, is called soft water. Water, which does not produce ready lather with soap, is known as hard water.

15) What are temporary and permanent hardness.?

The temporary hardness is due to the presence of bicarbonates of calcium and magnesium. Permanent hardness or non-carbonate hardness is caused by the presence of chlorides and sulphates of calcium, magnesium, aluminum etc.

16) Define ppm.

Parts per million(ppm) is the number of parts by weight of calcium carbonate equivalent hardness per million (10^6) parts of water

17) What is Break point chlorination?

The point at which free residual chlorine again begins to increase is the break point.

18) Define desalination.

The process of removing the common salt from the water is known as desalination.

19) What is carbonate conditioning?

Scale formation can be avoided by adding Na_2CO_3 to the boiler water. It is used only in low pressure boilers. Thescale forming salt like $CaSO_4$ is converted into $CaCO_3$, which can be removed easily

20) What are the advantages of RO method?

1. High life time.

2. Removes ionic, non-ionic and colloidal silica impurities , which cannot be removed bydemineralization method.

3. Low capital cost.

Part-B

1) Explain in detail about the food preservative.

2) Explain the chemical characteristics of water.

3) Explain the determination of hardness by EDTA method.

4) Explain in detail about domestic water treatment.

5) Describe the Demineralisation process with a neat diagram.

6) What are Internal conditioning methods? Explain in detail.

7) Explain the Reverse Osmosis method.

Unit II-Chemistry for Engineering Plastics

Part-A

1) How is nylon 6,6 formed?

It is a reaction between simple polar groups containing monomers with the formation of polymer and elimination of small molecules like $H_2O.HCl$ In order to form a condensation polymer the monomer must have two functional groups.

2) What are plastics?

The term plastics or plastics material in general is given to organic materials of high molecular weight, which can be moulded into any desired shape by the application heat and pressure in the presence of a catalyst.

3) What are the disadvantages of plastics?

- Very high softness, Embrittlement at low temperature.
- Deformation under load, Easily combustible.
- Low heat resistance.

4) List out the various ways by which polymers can be classified?

- Homo polymer.
- Hetero polymer (or) Co polymer.
- Homo chain polymer and Hetero chain polymer.

5) Why Teflon is behaving non sticky?

Since the fluorine atoms are the strong electro negative element they tightly bonds with carbon atoms in Teflon.As (C-F) bond is stronger it is non reactive, is not wetted by oil and H_2O. so Teflon is non sticky .

6) What are elastomers?

Rubbers or elastomers are noncrystalline high polymers (liner polymers) having elastic and other rubber like properties.

7) What is meant by vulcanization of rubber?

The process of vulcanization consists of heating the raw rubber with sulphur to about 100-140°C The added sulphur combines chemically at the double bond of different long chain rubber springs. Vulcanization prevents the intermolecular movement of rubber springs.

8) What is GR-I rubber

Butyl rubber or GR-I rubber is the copolymer of isobutylene and a small amount of isoprene.

9)   Why thermosetting plastics cannot be remolded

They are linked by strong covalent bonds.They consist of a 3d network structure.

10) What are the drawbacks of raw rubber?

It is plastic in nature i.e. it becomes soft at high temperature and is too brittle at low temperature.

- It has poor strength like 200Kg/cm2 at low temperature.
- It has a large water absorption capacity.
- It is non resistant to non polar solvents like benzene, vegetable and mineral oils.
- It is attacked by oxidizing agents like HNO_3 and H_2SO_4.

Part–B

1)   Write the free radical mechanism of polymerization.
2)   Explain molding of plastics with a neat diagram.
3)   Give the properties, preparation and uses of Teflon and Nylon 6,6.
4)   Explain why natural rubber needs vulcanization.
5)   In detail explain injection and blow molding technique.
6)   Differentiate thermoplastics and thermosetting plastics.
7)   Define polymerization reaction and explain the addition polymerization with suitable example.

Unit III–Electro Chemistry

Part-A

1) What is emf series?

The arrangement of standard reduction potential values in an orderly manner with reference to hydrogen electrode is called emf series.

2) What is an electrochemical cell?

Galvanic cells are electrochemical cells in which the electrons, transferred due to redox reaction, converted to electrical energy.

3) Define single electrode potential.

It is the measure of the tendency of a metallic electrode to lose or gain electrons, when it is in contact with a solution of its own salt.

4) Define electrolytes.What are the types of electrolytes?

Electrolyte is a water soluble substance forming ions in solution, and conduct an electric current. Types are Strong, weak and non- electrolyte.

5) What are reversible cells.

A cell which obey the following three conditions of thermodynamicreversiblity is called reversible cell.

6) What are ion selective electrodes? Give example.

Ion-selective electrodes are the electrodes having the ability to respond only to particular ions and develop potential, ignoring the other ions in a mixture totally. The potential developed by an ion-selective electrode depends only on the concentration of particular ions.

7) Define standard electrode potential.

It is the measure of the tendency of a metallic electrode to lose or gain electrons, when it is in contact with a solutionof its own salt of 1 molar concentration at 25°C.

8) What are the applications of Nernst equation.

- Nernst equation is used to calculate the electrode potential of unknown metal.
- The corrosion tendency of metals can be predicted.

9) What is corrosion?

The phenomenon of deterioration or destruction of metals and alloysis known as corrosion.

10) State Pilling –Bed worth rule.

The ratio between the volume of metal oxide to the volume of the metal is called Pilling-Bedworth rule. According to Pilling-Bedworth rule, if the volume of the oxide layer formed is less than the volume of the metal, the oxide layer is porous and non-protective. On the other hand, if the volume of the oxide layer formed is greater than thevolume of the metal, the oxide layer is non-porous and protective.

11) Write the advantages of electro less plating.

- No electricity is required.
- Electroless plating on insulators and semiconductors can be easily carried out.
- Complicated parts can also be plated uniformly.
- Electroless coating posses good mechanical, chemical and magnetic properties.

12) Define Electroplating.

The basic principle of electroplating is coating the coating material on the base metal by passing a direct current through an electrolytic solution containing the soluble salt of the coating material.

13) Define Electrolessplating

Electroless plating is a technique of depositing a noble metal (from its salt solution) on a catalytically active surface of the metal, to be protected , by using a suitable reducing agent without using electrical energy.

Part-B

1) What is reference electrode? Explain Calomel electrode with a neat diagram.

2) Derive Nernst equation.

3) Explain the working function of ion selective electrode with an example and its construction.

4) Define emf. How it is measured potentiometrically. Explain.

5) Discuss the mechanism of electrochemical corrosion with its types.

6) What is electro less plating? How is electro less plating of nickel carried out.

7) Explain the mechanism of dry corrosion.

8) Discuss the principle and methodology of gold plating.

UNIT IV- NON-CONVENTIONAL ENERGY SOURCES

Part-A

1) Differentiate between renewable and non renewable energy resources.

Renewable energy sources are inexhaustible & can be regenerated within a short span of time. But non renewable resources are exhausted & cannot be regenerated.

2) Differentiate between nuclear fission and fusion.

S. No	Nuclear fission	Nuclear fusion
1.	It is the process of breaking a heaviernucleus.	It is the process of combing alighter nucleus.
2	It emits radioactive rays	It does not emit any radioactive rays.
3.	It occurs at ordinary temperature.	It occurs at high temperature ($>10^6$K).
4.	The mass number and atomic numberof new elements are lower than that of parent nucleus.	The mass number, and atomic number ofproducts is higher than that of starting elements.
5.	It gives rise to the chain reaction.	It does not give rise to the chain reaction.
6.	It emits neutrons.	It emits positrons.
7.	It can be controlled.	It cannot be controlled.

3) What are moderators? Give an example.

The substances used to slow down the neutrons are called moderators

4) What are the types of solar energy conversions? Give an example.

Solar energy conversion is the process of conversion of direct sunlight into more useful forms. This solar energy conversion occurs by the following two mechanisms.

- Thermal conversion.
- Photoconversion

5) What is a fuel cell? Give an example.

The fuel cell is a voltaic cell, which converts the chemical energy of the fuels directly into electricity without combustion.It converts the energy of the fuel directly into electricity. In these cells, the reactants, products and electrolytes pass through the cell.Eg;Hydrogen-Oxygen fuel cell.

6) Lithium battery is considered as future cell. Why?

- Its cell voltage is high, 3.0 V.
- Since Li is a light-weight metal, only 7g (1 mole) material is required to produce 1 mole of electrons.
- Since Li has the most negative $E°$ value, it generates a higher voltage than the other types of cells.

- Since all the constituents of the battery are solids there is no risk of leakage from the battery.
- This battery can be made in a variety of sizes and shapes.

7) Give the advantages of hydrogen-oxygen fuel cell.

- Fuel cells are efficient (75%) and take less time for operation.
- It is pollution free technique.
- It produces electric current directly from the reaction.
- A fuel and an oxidiser.
- It produces drinking water.

8) Differentiate between primary and secondary batteries. Give examples.

- In these cells, the electrode and the electrode reactions cannot be reversed by passing an external electrical energy. The reactions occur only once and after use they become dead. Therefore, they are not chargeable.
- In these cells, the electrode reactions can be reversed by passing an external electrical energy. Therefore, they can be recharged by passing electric current and used again and again. These are also called Storage cells (or) Accumulators.

9) What are solar cells?

Photogalvanic cell is the one, which converts the solar energy (energy obtained from the sun) directly into electrical energy.

10) What are the disadvantages of wind energy?

- Public resists for locating the wind forms in populated areas due to noise generated by the machines and loss of aesthetic appearance.
- Wind forms located on the migratory routes of birds will cause hazards.
- Wind forms produce unwanted sound.
- Wind turbines interfere with electromagnetic signals (TV, Radio signals).

Part-B

1) Explain light water nuclear reactor with a neat diagram.
2) What are solar cells? Explain the applications of solar cells. Give its limitations.
3) Explain lead acid battery with neat diagram. Write the equations for charging and discharging.
4) Explain Nickel cadmium battery with neat diagram.
5) Write a note on fuel cells. Explain hydrogen oxygen fuel cell with neat diagram.

Unit V- Analytical Techniques

Part-A

1) Mention any four limitations of Beer Lambert's law.

Beer-Lambert's law is not obeyed if the radiation used is not monochromatic.

The temperature of the system should notbe allowed to vary to a large extent.

It is not applied to suspensions.

2) Mention any four applications of UV visible spectroscopy.

- Predicting relationship between different groups.
- Qualitative analysis.
- Quantitative analysis.

3) What is a finger print region? Give its uses.

The vibrational spectral (IR spectra) region at $1400 - 700$ cm-1 gives very rich and intense absorption bands.This region is termed as fingerprint region.

4) Mention any four applications of IR spectroscopy.

- Detection of functional groups.
- Qualitative analysis.
- Quantitative analysis.
- Study of progress of the reaction.

5) What is absorption spectrum?

When a beam of electromagnetic radiation is allowed to fall on a molecule in the ground state, the molecule absorbs photon of energy hv and undergoes a transition from the lower energy level to the higher energy level. The spectrum thus obtained is called the absorption spectrum

6) Define Beer-Lambert law.

According to this law, "when a beam of monochromatic radiation is passed through a solution of an absorbing substance, then rate of decrease of intensity of radiation 'dI' with thickness of the absorbing solution 'dx' is proportional to the intensity of incident radiation 'I'as well as the concentration of the solution 'C'.

7) What is nebulisation of liquid sample?

Before the liquid sample enters into the burner, it is first of all converted into liquid droplets. This is called nebulisation.

8) What are chromophores and auxochromes? Give an example.

The presence of one or more unsaturated linkages in a compound is responsible for the colour of the compound, these linkages are referred to as chromophores.Eg: -C=C,-N=N-

It refers to an atom or a group of atoms which does not give rise to absorption band on its own, but when conjugate to chromophore will cause a red shift.Eg:-OH, -Cl.

Part-B

1) Derive the equation for Beer Lambert's law.
2) Explain the instrumentation technique of colorimetric analysis.
3) Determine the   concentration of iron using colorimetric analysis
4) Explain the principle and instrumentation of UV visible spectroscopy
5) Explain the principle and instrumentation of IR spectroscopy
6) Explain the principle and instrumentation of flame photometry. How
7) will you estimate   sodium by flame photometry
8) Explain the principle and instrumentation of Atomic Absorption spectroscopy.